MULTIVARIATE DATA ANALYSIS

ASTROPHYSICS AND SPACE SCIENCE LIBRARY

A SERIES OF BOOKS ON THE RECENT DEVELOPMENTS
OF SPACE SCIENCE AND OF GENERAL GEOPHYSICS AND ASTROPHYSICS
PUBLISHED IN CONNECTION WITH THE JOURNAL
SPACE SCIENCE REVIEWS

Editorial Board

R.L.F. BOYD, *University College, London, England*

W. B. BURTON, *Sterrewacht, Leiden, The Netherlands*

L. GOLDBERG, *Kitt Peak National Observatory, Tucson, Ariz., U.S.A.*

C. DE JAGER, *University of Utrecht, The Netherlands*

J. KLECZEK, *Czechoslovak Academy of Sciences, Ondřejov, Czechoslovakia*

Z. KOPAL, *University of Manchester, England*

R. LÜST, *European Space Agency, Paris, France*

L. I. SEDOV, *Academy of Sciences of the U.S.S.R., Moscow, U.S.S.R.*

Z. ŠVESTKA, *Laboratory for Space Research, Utrecht, The Netherlands*

MULTIVARIATE DATA ANALYSIS

by

FIONN MURTAGH

*ST-ECF/European Southern Observatory, Munich, F.R.G.
and Space Science Department, ESTEC, Noordwijk, The Netherlands*

and

ANDRÉ HECK

C.D.S., Observatoire Astronomique, Strasbourg, France

D. REIDEL PUBLISHING COMPANY

A MEMBER OF THE KLUWER  ACADEMIC PUBLISHERS GROUP

DORDRECHT / BOSTON / LANCASTER / TOKYO

Library of Congress Cataloging in Publication Data

Murtagh, Fionn.
  Multivariate data analysis

  (Astrophysics and space science library)
  Bibliography: p.
  Includes index.
  1. Statistical astronomy. 2. Multivariate analysis. I. Heck, A. (André) II. Title.
QB149.M87   1987   520′.1′519535   86–29821
ISBN 90–277–2425–3

Published by D. Reidel Publishing Company,
P.O. Box 17, 3300 AA Dordrecht, Holland.

Sold and distributed in the U.S.A. and Canada
by Kluwer Academic Publishers,
101 Philip Drive, Assinippi Park, Norwell, MA 02061, U.S.A.

In all other countries, sold and distributed
by Kluwer Academic Publishers Group,
P.O. Box 322, 3300 AH Dordrecht, Holland.

All Rights Reserved
© 1987 by D. Reidel Publishing Company, Dordrecht, Holland
No part of the material protected by this copyright notice may be reproduced or
utilized in any form or by any means, electronic or mechanical
including photocopying, recording or by any information storage and
retrieval system, without written permission from the copyright owner

Printed in The Netherlands

To Gladys and Sorcha.
To Marianne.

Contents

List of Figures — xi

List of Tables — xiii

Foreword — xv

1 Data Coding and Initial Treatment — 1
- 1.1 The Problem — 1
- 1.2 Mathematical Description — 2
  - 1.2.1 Introduction — 2
  - 1.2.2 Distances — 3
  - 1.2.3 Similarities and Dissimilarities — 6
- 1.3 Examples and Bibliography — 8
  - 1.3.1 Selection of Parameters — 8
  - 1.3.2 Example 1: Star–Galaxy Separation — 8
  - 1.3.3 Example 2: Galaxy Morphology Classification — 9
  - 1.3.4 General References — 10

2 Principal Components Analysis — 13
- 2.1 The Problem — 13
- 2.2 Mathematical Description — 15
  - 2.2.1 Introduction — 15
  - 2.2.2 Preliminaries — Scalar Product and Distance — 17
  - 2.2.3 The Basic Method — 18
  - 2.2.4 Dual Spaces and Data Reduction — 21
  - 2.2.5 Practical Aspects — 23
  - 2.2.6 Iterative Solution of Eigenvalue Equations — 25
- 2.3 Examples and Bibliography — 26
  - 2.3.1 General Remarks — 26
  - 2.3.2 Artificial Data — 27
  - 2.3.3 Examples from Astronomy — 29
  - 2.3.4 General References — 33

	2.4		Software and Sample Implementation	33
		2.4.1	Progam Listing	34
		2.4.2	Input Data	49
		2.4.3	Sample Output	51
3	**Cluster Analysis**			**55**
	3.1		The Problem	55
	3.2		Mathematical Description	56
		3.2.1	Introduction	56
		3.2.2	Hierarchical Methods	57
		3.2.3	Agglomerative Algorithms	61
		3.2.4	Minimum Variance Method in Perspective	67
		3.2.5	Minimum Variance Method: Mathematical Properties	70
		3.2.6	Minimal Spanning Tree	71
		3.2.7	Partitioning Methods	73
	3.3		Examples and Bibliography	77
		3.3.1	Examples from Astronomy	77
		3.3.2	General References	83
	3.4		Software and Sample Implementation	85
		3.4.1	Program Listing: Hierarchical Clustering	86
		3.4.2	Program Listing: Partitioning	99
		3.4.3	Input Data	106
		3.4.4	Sample Output	106
4	**Discriminant Analysis**			**111**
	4.1		The Problem	111
	4.2		Mathematical Description	113
		4.2.1	Multiple Discriminant Analysis	113
		4.2.2	Linear Discriminant Analysis	115
		4.2.3	Bayesian Discrimination: Quadratic Case	116
		4.2.4	Maximum Likelihood Discrimination	119
		4.2.5	Bayesian Equal Covariances Case	120
		4.2.6	Non–Parametric Discrimination	120
	4.3		Examples and Bibliography	122
		4.3.1	Practical Remarks	122
		4.3.2	Examples from Astronomy	123
		4.3.3	General References	125
	4.4		Software and Sample Implementation	127
		4.4.1	Program Listing: Linear Discriminant Analysis	128
		4.4.2	Program Listing: Multiple Discriminant Analysis	134
		4.4.3	Program Listing: K–NNs Discriminant Analysis	148
		4.4.4	Input Data	151

		4.4.5	Sample Output: Linear Discriminant Analysis	151
		4.4.6	Sample Output: Multiple Discriminant Analysis	152
		4.4.7	Sample Output: K-NNs Discriminant Analysis	154

5 Other Methods — 155

- 5.1 The Problems ... 155
- 5.2 Correspondence Analysis ... 156
  - 5.2.1 Introduction ... 156
  - 5.2.2 Properties of Correspondence Analysis ... 157
  - 5.2.3 The Basic Method ... 159
  - 5.2.4 Axes and Factors ... 159
  - 5.2.5 Multiple Correspondence Analysis ... 160
- 5.3 Principal Coordinates Analysis ... 163
  - 5.3.1 Description ... 163
  - 5.3.2 Multidimensional Scaling ... 166
- 5.4 Canonical Correlation Analysis ... 167
- 5.5 Regression Analysis ... 167
- 5.6 Examples and Bibliography ... 169
  - 5.6.1 Regression in Astronomy ... 169
  - 5.6.2 Regression in General ... 171
  - 5.6.3 Other Techniques ... 171

6 Case Study: IUE Low Dispersion Spectra — 173

- 6.1 Presentation ... 173
- 6.2 The IUE Satellite and its Data ... 173
- 6.3 The Astrophysical Context ... 174
- 6.4 Selection of the Sample ... 176
- 6.5 Definition of the Variables ... 176
  - 6.5.1 The Continuum Asymmetry Coefficient ... 177
  - 6.5.2 The Reddening Effect ... 178
- 6.6 Spectral Features ... 179
  - 6.6.1 Generalities ... 179
  - 6.6.2 Objective Detection of the Spectral Lines ... 180
  - 6.6.3 Line Intensities ... 182
  - 6.6.4 Weighting Line Intensities ... 182
- 6.7 Multivariate Analyses ... 182
  - 6.7.1 Principal Components Analysis ... 183
  - 6.7.2 Cluster Analysis ... 184
  - 6.7.3 Multiple Discriminant Analysis ... 191

7 Conclusion: Strategies for Analysing Data — 195

- 7.1 Objectives of Multivariate Methods ... 195

7.2	Types of Input Data	196
7.3	Strategies of Analysis	197

General References   **199**

Index   **205**

List of Figures

1.1 Two records (x and y) with three variables (Seyfert type, magnitude, X-ray emission) showing disjunctive coding. 5

2.1 Points and their projections onto axes. 14
2.2 Some examples of PCA of centred clouds of points. 16
2.3 Projection onto an axis. 19

3.1 Construction of a dendrogram by the single linkage method. ... 58
3.2 Differing representations of a hierarchic clustering on 6 objects. . . 60
3.3 Another approach to constructing a single linkage dendrogram. . . 62
3.4 Five points, showing *NN*s and *RNN*s. 66
3.5 Alternative representations of a hierarchy with an inversion. 67
3.6 Three binary hierarchies: symmetric, asymmetric and intermediate. 69
3.7 Minimal spanning tree of a point pattern (non-unique). 73
3.8 Differing point patterns. 74

4.1 Two sets of two groups. Linear separation performs adequately in (a) but non-linear separation is more appropriate in (b). 112
4.2 The assignment of a new sample **a** to one of two groups of centres $\mathbf{y}_1$ and $\mathbf{y}_2$. 116

5.1 Table in complete disjunctive form and associated Burt table. ... 161
5.2 Horseshoe pattern in principal plane of Correspondence Analysis. . 164

6.1 The International Ultraviolet Explorer (IUE). 174
6.2 Illustration of terms in the asymmetry coefficient S (see text). ... 178
6.3 Illustration of the reddening effect on a normalized spectrum. ... 179
6.4 Test used for objectively detecting potential lines in the spectra. . 180
6.5 Weighting of the line intensities by the asymmetry coefficient through the Variable Procrustean Bed technique. 183
6.6 Dwarf and supergiant stars in the plane of discriminant factors 1 and 2. 192

6.7 Giant and supergiant stars in the plane of discriminant factors 1 and 3. 193

List of Tables

3.1	Specifications of seven hierarchical clustering methods.	64
5.1	Properties of spaces $\mathbb{R}^m$ and $\mathbb{R}^n$ in Correspondence Analysis. . . .	158
6.1	UV spectral distribution of the stars sampled.	176
6.2	Bin wavelengths corresponding to the 60 most frequent lines in the spectral sample at hand.	181
6.3	Star identifiers, IUE spectral classification, and 30–, 40– and 50– cluster solutions.	190

Foreword

> You may ask: "What can a hard headed statistician offer to a starry eyed astronomer?" The answer is: "Plenty."
>
> *Narlikar (1982)*

In the past a physical scientist, and in particular an astronomer, worked individually, from the conception of a project through to the collection and analysis of data. As instrumentation became more complex, teams of researchers became necessary, and these teams progressively included specialists in technology (whether astronomers or otherwise). Today it is virtually impossible to run a project at the forefront of research in natural science without the help of these technologists.

At the other end of the chain, it can be seen too that in the future teams will have to include specialists in methodology to work on collected data. Image processing alone is not in question here (it is, of course, a natural consequence of sophisticated technology), but rather methodology applicable to already well-reduced data. This is the only way to face the challenge raised by the accumulation of data.

Compared to the past, ever larger amounts of data are indeed being collected, and the rate will continue to accelerate. The Space Telescope will send down, over an estimated lifetime of 15 years, the equivalent of 6×10^{12} bytes of information, or a daily average of 10^9 bytes of data. Even apart from the Space Telescope, there are now more and more instruments at the astronomer's disposal. The rate of increase of observations is also matched by the diversification of the data.

It will therefore be necessary to work on bigger samples if full advantage is to be taken of the information contained in such quantities of data. It will similarly be necessary to derive as much information as possible from the diversity of the data, rather than restricting attention to subsets of it. We might well live at the end of the period when a significant number of astronomers spend their lives investigating a couple of pet objects. If this is not the case, what is the use of collecting so much data?

One way to work effectively on large samples is to apply, and if necessary to develop, an adequate statistical methodology. If Nature is consistent, results obtained from the application of tools developed by mathematicians and statisticians should not be in contradiction with those obtained by physical analyses.

Statistical methods are not of course intended to replace physical analysis. These should be seen as complementary, and statistical methods can be used to run a rough preliminary investigation, to sort out ideas, to put a new ("objective" or "independent") light on a problem, or to point out aspects which would not come out in a classical approach. Physical analysis is necessary to subsequently refine and interpret the results, and to take care of the details.

Interest in the methods studied in the following chapters is increasing rapidly in the astronomical community, as shown for example by successful conferences and working groups in this area. It has been felt that the availability of introductory material in this area is long overdue. The aim of this book is to provide an introduction to a number of the more useful techniques in multivariate statistics.

A wide-ranging annotated set of bibliographic references follow each chapter, and allow further study in this area. These references should provide valuable entry-points for research-workers in differing astronomical sub-disciplines.

Although the applications considered in this book are focussed on astronomy, the algorithms presented can be applied to similar problems in other branches of science. FORTRAN-77 programs are provided for many of the methods described. For these programs a sample input data set is given, and the output obtained is reproduced so that it can be used in implementation-testing. The programs, together with others in this area, are also available in the European Southern Observatory's MIDAS (Munich Image Data Analysis System) system.

The authors are pleased to acknowledge that Professor P. Benvenuti (Head, Space Telescope — European Coordinating Facility) initially introduced them to each other a few years ago. Appreciation also goes to Professor C. Jaschek (Director, Strasbourg Data Centre) for encouraging the present work. Finally, a number of valuable comments on the manuscript (especially from Dr. F. Ochsenbein, European Southern Observatory) were very much appreciated. This book would not have been possible without the welcoming resources of the European Southern Observatory, including the use of the LaTeX document preparation system.

F. Murtagh and A. Heck,
Munich and Strasbourg.

Chapter 1

Data Coding and Initial Treatment

1.1 The Problem

Data Analysis has different connotations for different researchers. It is however definitely part of the chain of reduction and exploitation of data, located somewhere between, let us say, the taking of observations and the eventual, published article.

The versatility and range of possible applications of multivariate data analysis make it important to be understood by astronomical researchers. Its applicability-boundaries — or equally, the requirements for input data for such algorithms — should also be appreciated.

In the following chapters, the techniques to be studied are

Principal Components Analysis: which focusses on inter–object correlations, reduces their (parameter–space) dimensionality, and allows planar graphic representations of the data;

Cluster Analysis: which is the application of automatic grouping procedures;

Discriminant Analysis: which classifies items into pre–defined groups.

A number of other, related methods will also be studied.

The data array to be analysed crosses objects (e.g. ultraviolet spectra, spiral galaxies or stellar chemical abundances) with variables or parameters. The former are usually taken as the rows of the array, and the latter as the columns. The choice of parameters to use for the study of a particular set of objects is of vital importance, but concrete guidelines cannot unfortunately be given in advance. The results of the analysis will depend in large measure on the choice of parameters (and may indeed be used to judge the relevance of the set of parameters chosen).

The following remarks should be borne in mind, however, in choosing parameters to suitably characterise the objects under investigation.

- As far as possible the parameters should be *homogeneous*: in multivariate data analysis, we do not wish to find differences in the objects being explained purely due to inhomogeneities in the parameters used.

- The parameters should allow for a *comprehensive* description of the objects; i.e. express as many aspects of the objects as possible.

- They, and the objects, should be chosen so that there is no *bias* in what is being studied (Pfleiderer, 1983). There should also be sufficient objects to constitute a reasonable sample (as a very general rule, a few dozen).

- As far as possible, *missing data* should be avoided.

Later in this chapter, we will describe the selection of parameters in two short case–studies.

Unlike classical, inferential statistics, there is rarely need for distributional assumptions in the methods to be studied. Multivariate methods have *description* of the data as their aim, — and hence the drawing of conclusions. Less precision in the possible conclusions of multivariate methods is balanced by the greater range of situations where they can be applied. Nonetheless, some knowledge of classical statistics is a *sine qua non* for successful analysis of data. Wall (1979) or Wetherill (1972), among many other texts, provide introductory material in this area.

The object–parameter dependencies can take many forms. Most suitable from the point of view of many of the algorithms to be discussed in subsequent chapters is *quantitative* data, i.e. real valued data, positive or negative, defined relative to some zero point. Another form of data is *qualitative* or *categorical*, i.e. the object–parameter relation falls into one of a number of categories; in its most simple form, this is a yes/no dependency indicating the presence or absence of the parameter. The coding of a categorical variable or parameter may take the general form of values "a", "b", etc. for the different categories, or "1", "2", etc. (where in the latter case, the values have "qualitative" significance only). A final form of data to be mentioned here is *ordinal* data where a rank characterises the object on the parameter.

1.2 Mathematical Description

1.2.1 Introduction

In the great majority of multivariate data analysis methods, the notion of distance (or similarity) is central. In clustering, objects are clustered on the basis of their mutual similarity, for instance, and in Principal Components Analysis (PCA) the

1.2. MATHEMATICAL DESCRIPTION

points are considered as vectors in a metric space (i.e. a space with a distance defined).

A very large number of coefficients for measuring similarity or distance have at one time or another been proposed. We will not attempt an inventory in the following sections but will instead deal with commonly used coefficients.

If the data to be analysed is of conveniently small size, then a visual scan of pairwise distances can reveal interesting features. It could be asserted that descriptive procedures such as PCA and clustering provide means of exploring such distances when it becomes impractical to do it "manually".

Some of the problems which arise in deciding on a suitable distance are as follows. If the data to be analysed is all of one type, a suitable distance can be chosen without undue difficulty. If the values on different coordinates are quite different — as is more often the case — some *scaling* of the data will be required before using a distance. Equally, we may view the use of a distance on scaled data as the definition of another, new distance. More problemsome is the case of data which is of mixed type (e.g. quantitative, categorical, ordinal values). Here, it may be possible to define a distance which will allow all coordinates to be simultaneously considered. It may be recommendable, though, to redefine certain coordinate variables (e.g. to consider ordinal values as quantitative or real variables). As a general rule, it is usually best to attempt to keep "like with like". A final problem area relates to missing coordinate values: as far as possible care should be taken in the initial data collection phase to ensure that all values are present.

1.2.2 Distances

Proximity between any pair of items will be defined by *distance*, *dissimilarity* or *similarity*. *Distance* is simply a more restrictive *dissimilarity*, — it satisfies certain axioms listed below. Both *distances* and *dissimilarities* measure identical items by a zero value, and by increasingly large (positive) values as the proximity of the items decreases. *Similarities* are mathematical functions which treat pairs of items from the other perspective: large values indicate large proximity, while small (positive) or zero values indicate little or no proximity. The mathematician is happiest when dealing with *distances*: an established body of theory is immediately available, and many of the methods to be studied in subsequent chapters work on distances.

The most commonly used distance for quantitative (or continuous) data is the *Euclidean* distance. If

$$\mathbf{a} = \{a_j : j = 1, 2, ..., m\}$$

and
$$\mathbf{b} = \{b_j : j = 1, 2, ..., m\}$$
are two real-valued vectors then the unweighted squared Euclidean distance is given by

$$d^2(a,b) = \sum_j (a_j - b_j)^2 = (\mathbf{a} - \mathbf{b})'(\mathbf{a} - \mathbf{b})$$

where **a** and **b** are taken as column vectors, and $'$ denotes transpose,

$$= \|\mathbf{a}\|^2 + \|\mathbf{b}\|^2 - 2\mathbf{a}'\mathbf{b}$$

where $\|.\|$ is the norm, or distance from the origin.

The Euclidean, along with all distances, satisfies the following properties:

Symmetry: $d(a,b) = d(b,a)$.

Positive semi-definiteness: $d(a,b) > 0$, if $a \neq b$; $d(a,b) = 0$, if $a = b$.

Triangular inequality: $d(a,b) \leq d(a,c) + d(c,b)$.

The triangular inequality in the Euclidean plane is simply the assertion that going from a to b cannot be shortened by going first to another point c; equality corresponds to point c being located on the line connecting a and b. Some further aspects of the Euclidean distance are discussed in Chapter 2.

The *Hamming* distance is an alternative to the Euclidean distance:

$$d(a,b) = \sum_j |a_j - b_j|$$

where $|.|$ is absolute value.

When *binary* data is being studied (i.e. categorical data with presence/absence values for each object, on each parameter) mutual possession of a property contributes 0 to this distance, mutual non-possession also contributes 0, and presence/absence contributes 1.

The Hamming and Euclidean distances are both members of a general class of distances termed the *Minkowski* metrics:

$$d_p(a,b) = \sqrt[p]{\sum_j |a_j - b_j|^p} \qquad p \geq 1.$$

When $p = 2$ we have the Euclidean distance; $p = 1$ gives the Hamming distance; and $p = \infty$ reduces to

$$d_\infty(a,b) = max_j |a_j - b_j|$$

1.2. MATHEMATICAL DESCRIPTION

	Record x:	S1, 18.2, X
	Record y:	S1, 6.7, —

	Seyfert type spectrum				Integrated magnitude		X-ray data?
	S1	S2	S3	—	≤ 10	> 10	Yes
x	1	0	0	0	0	1	1
y	1	0	0	0	1	0	0

Figure 1.1: Two records (x and y) with three variables (Seyfert type, magnitude, X-ray emission) showing disjunctive coding.

which is the "maximum coordinate" or *Chebyshev* distance. These three are the most commonly used Minkowski distances. The corresponding norms, i.e.

$$\|\mathbf{a}\|_p = d_p(a, 0)$$

where 0 is the origin, are referred to as the L_1, L_2, and L_∞ norms.

Since these distances are symmetric, it is not necessary to store all n interpair distances for n objects: $n(n-1)/2$ suffice.

With a suitable coding, the Euclidean, Hamming or other distances may be used in a wide variety of circumstances. Consider in Figure 1.1 part of a set of records from a Quasar catalogue. Part of two records, x and y, are shown. In order to carry out a comparison of such records, we can as a preliminary recode each variable, as shown. The Hamming and Euclidean distances are both then

$$d(x, y) = 0 + 0 + 0 + 0 + 1 + 1 + 1 = 3$$

in this case. The breakdown of Seyfert spectrum type in this example is easier to accomplish than the breakdown of positional coordinates (it would be necessary to astutely break up the angle values into useful, but ad-hoc, categories). User choice is required in defining such a disjunctive form of coding. In the case of quantitative variables this coding may be especially useful when one is in the presence of widely varying values: specifying a set of categories may make the distances less vulnerable to undesirable fluctuations.

The possible preponderance of certain variables in, for example, a Euclidean distance leads to the need for a *scaling* of the variables, i.e. for their *centring* (zero mean), *normalization* (unit variance), or *standardization* (zero mean and unit standard deviation). If a_{ij} is the jth coordinate of vector $\mathbf{a}_i$ (i.e. the ij^{th} table entry), $\bar{a}_j$ is the mean value on coordinate (variable) j, and σ_j is the standard deviation of variable j, then we standardize a_{ij} by transforming it to

$$(a_{ij} - \bar{a}_j)/\sigma_j$$

where

$$\bar{a}_j = \sum_{i=1}^{n} a_{ij}/n$$

$$\sigma_j^2 = \sum_{i=1}^{n} (a_{ij} - \bar{a}_j)^2/n$$

and n is the number of rows in the given table.

Standardization, defined in this way, is widely used, but nothing prevents some alternative scaling being decided on: for instance we could divide each value in the data matrix by the row mean, which has the subsequent effect of giving a zero distance to row vectors each of whose elements are a constant times the other. We may regard the resultant distance as a weighted Euclidean distance of the following form:

$$d^2(a, b) = \sum_j (w_1 a_j - w_2 b_j)^2.$$

Missing values constitute a current research problem: the simplest option is to delete rows or columns of the input data table in order to avoid these. There is no clear answer as to how to estimate missing values. However a dissimilarity, similar in form to a weighted Euclidean distance, may always be used for quantitative data:

$$d^2(a, b) = \sum_{j=1}^{m'} \frac{m'}{m} (a_j - b_j)^2$$

where $m' < m$ pairs of coordinates are simultaneously present for both **a** and **b**.

1.2.3 Similarities and Dissimilarities

A *dissimilarity* may satisfy some of the properties of a distance but often the triangular inequality is found to be the most difficult to satisfy.

Similarities, just as distances or dissimilarities, are functions mapping pairs of items (most often, vectors in a multidimensional space) into positive, real values.

In the case of binary data (i.e. data representing by 0 and 1 whether or not an object possesses a property), many similarity coefficients have been proposed. A similarity, s_{ij}, may be converted into a dissimilarity, δ_{ij}, when s_{max} is the maximum similarity present, in the following way:

$$\delta_{ij} = s_{max} - s_{ij}.$$

1.2. MATHEMATICAL DESCRIPTION

Similarity coefficients are often defined so that their resultant value will be between 0 and 1. For example, the Jaccard coefficient is defined for binary vectors **a** and **b**, and where N represents the counting operator, as

$$s(a,b) = \frac{N_j(a_j = b_j = 1)}{N_j(a_j = 1) + N_j(b_j = 1) - N_j(a_j = b_j = 1)}.$$

The numerator here reads: the number of times property j is simultaneously present for objects a and b. This can be written in vector notation as

$$s(a,b) = \frac{\mathbf{a'b}}{\mathbf{a'a} + \mathbf{b'b} - \mathbf{a'b}}.$$

As an example, the Jaccard similarity coefficient of vectors (10001001111) and (10101010111) is $5/(6+7-5) = 5/8$.

In the case of mixed quantitative–qualitative data, a suitable similarity coefficient may be difficult to define. The Gower coefficient considers all binary, other qualitative, and quantitative/ordinal coordinates in turn: for the binary, the Jaccard coefficient is determined; for other qualitative, a 1 indicates the same coordinate value, and a 0 indicates different coordinate values; and for the jth quantitative or ordinal coordinate of objects **a** and **b**, we determine

$$1 - \frac{|a_j - b_j|}{max_j |a_j - b_j|}.$$

The denominator is the maximum spread. It is seen that all contributions to the Gower similarity coefficient are between 0 and 1. A similarity coefficient with values between 0 and 1 can be obtained by

$$s = \frac{n_1 s_1 + n_2 s_2 + n_3 s_3}{n_1 + n_2 + n_3}$$

where n_1 coordinates are binary and s_1 is the Jaccard similarity obtained; n_2 coordinates are otherwise qualitative and s_2 is their contribution to the similarity discussed above; and finally n_3 coordinates are quantitative or ordinal.

For further discussion of similarities and distances, Anderberg (1973) or Everitt (1980) may be consulted. For their relation to metric properties, Gower and Legendre (1986) can be recommended.

We will conclude this section by pointing to a very different approach to defining dissimilarity, which constitutes an interesting research direction. In the case of spectral matching, without knowledge of calibration, a best fit between any pair of spectra must first be obtained and on the basis of this a dissimilarity defined. A technique from the Operations Research field, known as *dynamic programming*, may be used for such optimal matching. Dynamic programming offers a flexible approach for optimally matching two ordered sets of values. Further reading is to be found in Kruskal (1983), Sankoff and Kruskal (1983) and Hall and Dowling (1980).

1.3 Examples and Bibliography

1.3.1 Selection of Parameters

The annotated bibliographies of subsequent chapters may be referred to for studies in a wide range of areas, and where details of data coding may be found. Here, we will restrict ourselves to a few examples of the way in which parameter selection is carried out.

1.3.2 Example 1: Star–Galaxy Separation

In the case of star–galaxy classification, following the scanning of digitised images, Kurtz (1983) lists the following parameters which have been used:

1. mean surface brightness;
2. maximum intensity, area;
3. maximum intensity, intensity gradient;
4. normalized density gradient;
5. areal profile;
6. radial profile;
7. maximum intensity, 2nd and 4th order moments, ellipticity;
8. the fit of galaxy and star models;
9. contrast versus smoothness ratio;
10. the fit of a Gaussian model;
11. moment invariants;
12. standard deviation of brightness;
13. 2nd order moment;
14. inverse effective squared radius;
15. maximum intensity, intensity weighted radius;
16. 2nd and 3rd order moments, number of local maxima, maximum intensity.

1.3. EXAMPLES AND BIBLIOGRAPHY

References for all of these may be found in the cited work. Clearly there is room for differing views on parameters to be chosen for what is essentially the same problem! It is of course the case also that aspects such as the following will help to orientate us towards a particular set of parameters in a particular case: the quality of the data; the computational ease of measuring certain parameters; the relevance and importance of the parameters measured relative to the data analysis output (e.g. the classification, or the planar graphics); similarly, the importance of the parameters relative to theoretical models under investigation; and in the case of very large sets of data, storage considerations may preclude extravagance in the number of parameters used.

1.3.3 Example 2: Galaxy Morphology Classification

The generating of a galaxy equivalent of the Hertzsprung–Russell diagram for stars, or the determining of a range of Hubble–type classes, using quantitative data on the galaxies, is a slowly burgeoning research topic. The inherent difficulty of characterising spirals (especially when not face–on) has meant that most work focusses on ellipticity in the galaxies under study. This points to an inherent bias in the potential multivariate statistical procedures. The inherent noisiness of the images (especially for faint objects) has additionally meant that the parameters measured ought to be made as robust as is computationally feasible. In the following, it will not be attempted to address problems of galaxy photometry per se (see Davoust and Pence, 1982; Pence and Davoust, 1985), but rather to draw some conclusions from recent (and ongoing) work which has the above-mentioned objectives in view.

From the point of view of multivariate statistical algorithms, a reasonably homogeneous set of parameters is required. Given this fact, and the available literature on quantitative galaxy morphological classification, two approaches to parameter selection appear to be strongly represented, albeit intermixed:

1. The luminosity profile along the major axis of the object is determined at discrete intervals. This may be done by the fitting of elliptical contours, followed by the integrating of light in elliptical annuli (Lefèvre *et al.*, 1986). A similar approach is used for the comprehensive database, currently being built up for the European Southern Observatory galaxy survey (by A. Lauberts). Noisiness and faintness require attention to robustness in measurement: the radial profile may be determined taking into account the assumption of a face–on optically–thin axisymmetric galaxy, and may be further adjusted to yield values for circles of given radius (Watanabe *et al.*, 1982). Alternatively, isophotal contours may determine the discrete radial values for which the profile is determined (Thonnat, 1985).

2. Specific morphology–related parameters may be derived instead of the profile. The integrated magnitude within the limiting surface brightness of 25 or 26 mag. arcsec^{-2} in the visual is popular (Takase et al., 1984; Lefèvre et al., 1986). The logarithmic diameter (D_{26}) is also supported by Okamura (1985). It may be interesting to fit to galaxies under consideration model bulges and disks using, respectively, $r^{\frac{1}{4}}$ or exponential laws (Thonnat, 1985), in order to define further parameters. Some catering for the asymmetry of spirals may be carried out by decomposing the object into octants; furthermore the taking of a Fourier transform of the intensity may indicate aspects of the spiral structure (Takase et al., 1984).

In the absence of specific requirements for the multivariate analysis, the following remarks can be made.

- The range of parameters to be used should be linked, if feasible, to the further use to which they might be put (i.e. to underlying theoretical aspects of interest).

- It appears that many parameters can be derived from a carefully–constructed luminosity profile, rather than it being possible to derive a profile from any given set of parameters. The utility of profiles would appear to be thereby favoured in the long term in this field.

- The presence of both types of data in the database is of course not a hindrance to analysis: however it is more useful if the analysis is carried out on both types of data separately.

Parameter data may be analysed by clustering algorithms, by Principal Components Analysis or by methods for Discriminant Analysis (all of which will be studied in the following chapters). Profile data may be sampled at suitable intervals and thus analysed also by the foregoing procedures. It may be more convenient in practice to create dissimilarities between profiles, and analyse these dissimilarities: this may be done using clustering algorithms with dissimilarity input.

1.3.4 General References

1. M.R. Anderberg, *Cluster Analysis for Applications*, Academic Press, New York, 1973.

2. E. Davoust and W.D. Pence, "Detailed bibliography on the surface photometry of galaxies", *Astronomy and Astrophysics Supplement Series*, **49**, 631–661, 1982.

3. B. Everitt, *Cluster Analysis*, Heinemann Educational Books, London, 1980 (2nd ed.).

4. J.C. Gower and P. Legendre, "Metric and Euclidean properties of dissimilarity coefficients" *Journal of Classification* **3**, 5–48, 1986.

5. P.A.V. Hall and G.R. Dowling, "Approximate string matching", *Computing Surveys*, **12**, 381–402, 1980.

6. C. Jaschek, "Data growth in astronomy", *Quarterly Journal of the Royal Astronomical Society*, **19**, 269–276, 1978.

7. J.B. Kruskal, "An overview of sequence comparison: time warps, string edits, and macromolecules", *SIAM Review*, **25**, 201–237, 1983.

8. M.J. Kurtz, "Classification methods: an introductory survey", in *Statistical Methods in Astronomy*, European Space Agency Special Publication 201, 47–58, 1983.

9. O. Lefèvre, A. Bijaoui, G. Mathez, J.P. Picat and G. Lelièvre, "Electronographic BV photometry of three distant clusters of galaxies", *Astronomy and Astrophysics*, **154**, 92–99, 1986.

10. J.V. Narlikar, "Statistical techniques in astronomy", *Sankhya: The Indian Journal of Statistics, Series B, Part 2*, **44**, 125–134, 1982.

11. S. Okamura, "Global structure of Virgo cluster galaxies", in eds. O.G. Richter and B. Binggeli, *ESO Workshop on The Virgo Cluster of Galaxies*, ESO Conference and Workshop Proceedings No. 20, 201–215, 1985.

12. W.D. Pence and E. Davoust, "Supplement to the detailed bibliography on the surface photometry of galaxies", *Astronomy and Astrophysics Supplement Series*, **60**, 517–526, 1985.

13. J. Pfleiderer, "Data set description and bias", in *Statistical Methods in Astronomy*, European Space Agency Special Publication 201, 3–11, 1983.

14. D. Sankoff and J.B. Kruskal (Eds.), *Time Warps, String Edits, and Macromolecules: The Theory and Practice of Sequence Comparison*, Addison-Wesley, New York, 1983.

15. B. Takase, K. Kodaira and S. Okamura, *An Atlas of Selected Galaxies*, University of Tokyo Press, Tokyo, 1984.

16. M. Thonnat, "Automatic morphological description of galaxies and classification by an expert system", INRIA Rapport de Recherche, Centre Sophia Antipolis, No. 387, 1985.

17. J.V. Wall, "Practical statistics for astronomers. I. Definitions, the normal distribution, detection of signal", *Quarterly Journal of the Royal Astronomical Society*, **20**, 130–152, 1979.

18. M. Watanabe, K. Kodaira and S. Okamura, "Digital surface photometry of galaxies toward a quantitative classification. I. 20 galaxies in the Virgo cluster", *Astrophysics Journal Supplement Series*, **50**, 1–22, 1982.

19. G.B. Wetherill, *Elementary Statistical Methods*, Chapman and Hall, London, 1972.

Chapter 2

Principal Components Analysis

2.1 The Problem

We have seen in Chapter 1 how the $n \times m$ data array which is to be analysed may be viewed immediately as a set of n row–vectors, or alternatively as a set of m column–vectors. PCA seeks the best, followed by successively less good, summarizations of this data. Cluster Analysis, as will be seen in Chapter 3, seeks groupings of the objects or attributes. By focussing attention on particular groupings, Cluster Analysis can furnish a more economic presentation of the data. PCA (and other techniques, as will be seen in a later chapter) has this same objective but a very different summarization of the data is aimed at.

In Figure 2.1a, three points are located in $\mathbb{R}^2$. We can investigate this data by means of the coordinates on the axes, taken separately. We might note, for instance, that on axis 1 the points are fairly regularly laid out (with coordinates 1, 2 and 3), whereas on axis 2 it appears that the points with projections 4 and 5 are somewhat separated from the point with projection 2. In higher–dimensional spaces we are limited to being easily able to visualize one–dimensional and two–dimensional representations (axes and planes), although at the limit we can construct a three–dimensional representation.

Given, for example, the array of 4 objects by 5 attributes,

$$\begin{pmatrix} 7 & 3 & 4 & 1 & 6 \\ 3 & 4 & 7 & 2 & 0 \\ 1 & 7 & 3 & -1 & 4 \\ 2 & 0 & -6 & 4 & 1 \end{pmatrix}$$

the projections of the 4 objects onto the plane constituted by axes 1 and 3 is simply

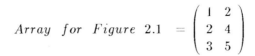

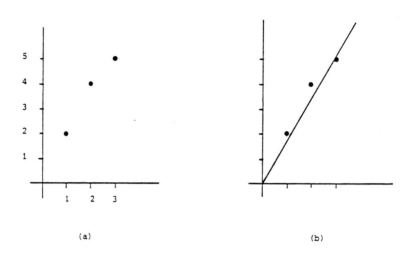

Figure 2.1: Points and their projections onto axes.

$$\begin{pmatrix} 7 & 4 \\ 3 & 7 \\ 1 & 3 \\ 2 & -6 \end{pmatrix}$$

Thus far, the projection of points onto axes or planes is a trivial operation. PCA, however, first obtains *better* axes. Consider Figure 2.1b where a new axis has been drawn as nearly as possible through all points. It is clear that if this axis went precisely through all points, then a second axis would be redundant in defining the locations of the points; i.e. the cloud of three points would be seen to be one–dimensional.

PCA seeks the axis which the cloud of points are closest to (usually the Euclidean distance defines *closeness*). This criterion will be seen below to be identical to another criterion: that the projections of points on the axis sought for be as elongated as possible. This second criterion is that the *variance* of the projections be as great as possible.

If, in general, the points under examination are m–dimensional, it will be very rare in practice to find that they approximately lie on a one–dimensional surface (i.e. a line). A second best–fitting axis, orthogonal to the first already found, will together constitute a best–fitting plane. Then a third best–fitting axis, orthogonal to the two already obtained, will together constitute a best–fitting three–dimensional subspace.

Let us take a few simple examples in two–dimensional space (Figure 2.2). Consider the case where the points are *centred* (i.e. the origin is located at the centre of gravity): this will usually be the case if the data are initially transformed to bring it about (see the previous Chapter and Section 2.2.5 below). We will seek the best–fitting axis, and then the next best–fitting axis. Figure 2.2a consists of just two points, which if centred must lie on a one–dimensional axis. In Figure 2.2b, the points are arranged at the vertices of a triangle. The vertical axis, here, accounts for the greatest variance, and the symmetry of the problem necessitates the positioning of this axis as shown. In the examples of Figure 2.2, the positive and negative orientations of the axes are arbitrary since they are not integral to our objective in PCA.

Central to the results of a PCA are the coordinates of the points (i.e. their projections) on the derived axes. These axes are listed in decreasing order of importance, or best–fit. Planar representations can also be output: the projections of points on the plane formed by the first and second new axes; then the plane formed by the first and third new axes; and so on, in accordance with user–request.

It is not always easy to remedy the difficulty of being unable to visualize high–dimensional spaces. Care must be taken when examining projections, since these may give a misleading view of interrelations among the points.

2.2 Mathematical Description

2.2.1 Introduction

The mathematical description of PCA which follows is important because other techniques (e.g. Discriminant Analysis) may be viewed as variants on it. It is one of the most straightforward geometric techniques, and is widely employed. These facts make PCA the best geometric technique to start with, and the most important to understand.

Section 2.2.2 will briefly deal with the basic notions of distance and projection to be used in subsequent sections. The Euclidean distance, in particular, has previously been studied in Chapter 1. Although some background knowledge of linear algebra is useful for the following sections, it is not absolutely necessary.

Section 2.2.3 will present the basic PCA technique. It answers the following questions:

- how is PCA formulated as a geometric problem ?

16   CHAPTER 2. PRINCIPAL COMPONENTS ANALYSIS

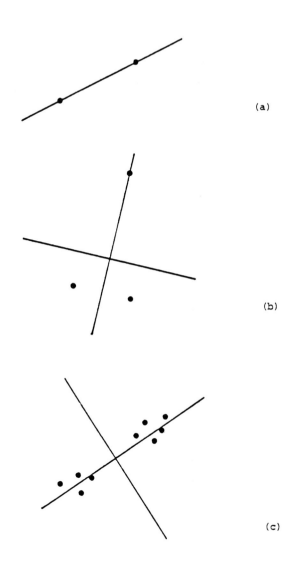

Figure 2.2: Some examples of PCA of centred clouds of points.

Such a quadratic form would increase indefinitely if **u** were arbitrarily large, so **u** is chosen — arbitrarily but reasonably — to be of unit length, i.e. $\mathbf{u'u} = 1$. We seek a maximum of the quadratic form $\mathbf{u'}S\mathbf{u}$ (where $S = X'X$) subject to the constraint that $\mathbf{u'u} = 1$. This is done by setting the derivative of the Lagrangian equal to zero. Differentiation of

$$\mathbf{u'}S\mathbf{u} - \lambda(\mathbf{u'u} - 1)$$

where λ is a Lagrange multiplier gives

$$2S\mathbf{u} - 2\lambda\mathbf{u}.$$

The optimal value of **u** (let us call it $\mathbf{u}_1$) is the solution of

$$S\mathbf{u} = \lambda\mathbf{u}.$$

The solution of this equation is well–known: **u** is the eigenvector associated with the eigenvalue λ of matrix S. Therefore the eigenvector of $X'X$, $\mathbf{u}_1$, is the axis sought, and the corresponding largest eigenvalue, λ_1, is a figure of merit for the axis, — it indicates the amount of variance explained by the axis (see next section).

The second axis is to be orthogonal to the first, i.e. $\mathbf{u'u}_1 = 0$, and satisfies the equation

$$\mathbf{u'}X'X\mathbf{u} - \lambda_2(\mathbf{u'u} - 1) - \mu_2(\mathbf{u'u}_1)$$

where λ_2 and μ_2 are Lagrange multipliers. Differentiating gives

$$2S\mathbf{u} - 2\lambda_2\mathbf{u} - \mu_2\mathbf{u}_1.$$

This term is set equal to zero. Multiplying across by $\mathbf{u}_1'$ implies that μ_2 must equal 0. Therefore the optimal value of **u**, $\mathbf{u}_2$, arises as another solution of $S\mathbf{u} = \lambda\mathbf{u}$. Thus λ_2 and $\mathbf{u}_2$ are the second largest eigenvalue and associated eigenvector of S.

The eigenvectors of $S = X'X$, arranged in decreasing order of corresponding eigenvalues, give the line of best fit to the cloud of points, the plane of best fit, the three–dimensional hyperplane of best fit, and so on for higher–dimensional subspaces of best fit. $X'X$ is referred to as the *sums of squares and cross products* matrix.

It has been assumed that the eigenvalues decrease in value: equal eigenvalues are possible, and indicate that equally privileged directions of elongation have been found. In practice, the set of equal eigenvalues may be arbitrarily ordered in any convenient fashion. Zero eigenvalues indicate that the space is actually of dimensionality less than expected (the points might, for instance, lie on a plane in three–dimensional space). The relevance of zero–valued eigenvalues is returned to in Section 2.3.1 below.

2.2. MATHEMATICAL DESCRIPTION

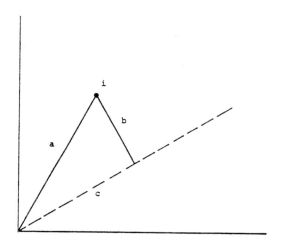

For each point i,

b = distance of i from new axis,

c = projection of vector i onto new axis,

a = distance of point from origin.

By Pythagoras, $a^2 = b^2 + c^2$; since a is constant, the choice of new axis which minimizes b simultaneously maximizes c (both of which are summed over all points, i).

Figure 2.3: Projection onto an axis.

set of objects and j a member of the attribute set. The objects may be regarded as row vectors in $\mathbb{R}^m$ and the attributes as column vectors in $\mathbb{R}^n$.

In $\mathbb{R}^m$, the space of objects, PCA searches for the best–fitting set of orthogonal axes to replace the initially–given set of m axes in this space. An analogous procedure is simultaneously carried out for the dual space, $\mathbb{R}^n$. First, the axis which best fits the objects/points in $\mathbb{R}^m$ is determined. If $\mathbf{u}$ is this vector, and is of unit length, then the product $X\mathbf{u}$ of $n \times m$ matrix by $m \times 1$ vector gives the projections of the n objects onto this axis.

The criterion of goodness of fit of this axis to the cloud of points will be defined as the squared deviation of the points from the axis. Minimizing the sum of distances between points and axis is equivalent to maximizing the sum of squared projections onto the axis (see Figure 2.3), i.e. to maximizing the variance (or *spread*) of the points when projected onto this axis.

The squared projections of points on the new axis, for all points, is

$$(X\mathbf{u})'(X\mathbf{u}).$$

$$\|\mathbf{x}\|^2 = \mathbf{x}'\mathbf{x}$$

which is easily verified to be > 0 if $\mathbf{x} \neq \mathbf{0}$ and $= 0$ otherwise; and secondly distance (squared):

$$\|\mathbf{x} - \mathbf{y}\|^2 = (\mathbf{x} - \mathbf{y})'(\mathbf{x} - \mathbf{y})$$

which is usually denoted by $d^2(x, y)$.

A general Euclidean space allows for non–orthogonal axes using the following definitions:

Scalar product: $\mathbf{x}'M\mathbf{y} = \mathbf{y}'M\mathbf{x}$ (M must be symmetric).

Norm (squared): $\|\mathbf{x}\|_M^2 = \mathbf{x}'M\mathbf{x}$ (M must be positive definite so that the norm is positive, or zero if $\mathbf{x} = \mathbf{0}$).

Distance (squared): $\|\mathbf{x} - \mathbf{y}\|_M^2 = (\mathbf{x} - \mathbf{y})'M(\mathbf{x} - \mathbf{y})$.

M–orthogonality: $\mathbf{x}$ is M–orthogonal to $\mathbf{y}$ if $\mathbf{x}'M\mathbf{y} = 0$.

Unless otherwise specified, the usual Euclidean space associated with the identity matrix is understood (i.e. M has ones on the main diagonal and zeros elsewhere).

The projection of vector $\mathbf{x}$ onto axis $\mathbf{u}$ is

$$\mathbf{y} = \frac{\mathbf{x}'M\mathbf{u}}{\|\mathbf{u}\|_M} \mathbf{u}$$

i.e. the coordinate of the projection on the axis is $\mathbf{x}'M\mathbf{u}/\|\mathbf{u}\|_M$. This becomes $\mathbf{x}'M\mathbf{u}$ when the vector $\mathbf{u}$ is of unit length.

The cosine of the angle between vectors $\mathbf{x}$ and $\mathbf{y}$ in the usual Euclidean space is $\mathbf{x}'\mathbf{y}/\|\mathbf{x}\|\|\mathbf{y}\|$. (That is to say, we make use of the triangle whose vertices are the origin, the projection of $\mathbf{x}$ onto $\mathbf{y}$, and vector $\mathbf{x}$. The cosine of the angle between $\mathbf{x}$ and $\mathbf{y}$ is then the coordinate of the projection of $\mathbf{x}$ onto $\mathbf{y}$, divided by the — hypotenuse — length of $\mathbf{x}$). The correlation coefficient between two vectors is then simply the cosine of the angle between them, when the vectors have first been centred (i.e. $\mathbf{x} - \mathbf{g}$ and $\mathbf{y} - \mathbf{g}$ are used, where $\mathbf{g}$ is the overall centre of gravity).

The notions of distance and of projection will be central in the description of PCA (to be looked at next) and in the description of other geometric data analytic techniques studied in subsequent chapters.

2.2.3 The Basic Method

Consider a set of n objects measured on each of m attributes or variables. The $n \times m$ matrix of values will be denoted by $X = \{x_{ij}\}$ where i is a member of the

2.2. MATHEMATICAL DESCRIPTION

- having defined certain geometric criteria, how is PCA formulated as an optimisation problem ?

- finally, how is PCA related to the eigen–decomposition of a matrix ?

Although a range of solution techniques for the eigenproblem are widely implemented in packages and subroutine libraries, an intuitively simple iterative solution technique is described later in Section 2.2.6.

Section 2.2.4 relates the use of PCA on the n (row) points in space $\mathbb{R}^m$ and the PCA of the m (column) points in space $\mathbb{R}^n$. Secondly, arising out of this mathematical relationship, the use of PCA as a data reduction technique is described. This section, then, answers the following questions:

- how is the PCA of an $n \times m$ matrix related to the PCA of the transposed $m \times n$ matrix ?

- how may the new axes derived — the principal components — be said to be linear combinations of the original axes ?

- how may PCA be understood as a series expansion ?

- in what sense does PCA provide a lower–dimensional approximation to the original data ?

In practice the variables on which the set of objects are characterised may often have differing means (consider the case of n students and m examination papers: the latter may be marked out of different totals, and furthermore it is to be expected that different examiners will set different standards of rigour). To circumvent this and similar difficulties, PCA is usually carried out on a transformed data matrix. Section 2.2.5 describes this transformation.

As an aid to the memory in the mathematics of the following sections, it is often convenient to note the dimensions of the vectors and matrices in use; it is vital of course that consistency in dimensions be maintained (e.g. if $\mathbf{u}$ is a vector in $\mathbb{R}^m$ of dimensions $m \times 1$, then premultiplication by the $n \times m$ matrix X will yield a vector $X\mathbf{u}$ of dimensions $n \times 1$).

2.2.2 Preliminaries — Scalar Product and Distance

This section defines a general Euclidean space. The usual Euclidean space is defined by means of its scalar product

$$\mathbf{x}'\mathbf{y} = \mathbf{y}'\mathbf{x}$$

(where $'$ denotes transpose). From the definition of scalar product, we may derive firstly the norm (here it is squared):

2.2.4 Dual Spaces and Data Reduction

In the dual space of attributes, $\mathbb{R}^n$, a PCA may equally well be carried out. For the line of best fit, $\mathbf{v}$, the following is maximized:

$$(X'\mathbf{v})'(X'\mathbf{v})$$

subject to

$$\mathbf{v}'\mathbf{v} = 1.$$

In $\mathbb{R}^m$ we arrived at

$$X'X\mathbf{u}_1 = \lambda_1 \mathbf{u}_1.$$

By similar reasoning, in $\mathbb{R}^n$, we have

$$XX'\mathbf{v}_1 = \mu_1 \mathbf{v}_1.$$

Premultiplying the first of these relationships by X yields

$$(XX')(X\mathbf{u}_1) = \lambda_1 (X\mathbf{u}_1)$$

and so $\lambda_1 = \mu_1$ (because we have now arrived at two eigenvalue equations which are identical in form). We must be a little more attentive to detail before drawing a conclusion on the relationship between the eigenvectors in the two spaces: in order that these be of unit length, it may be verified that

$$\mathbf{v}_1 = \frac{1}{\sqrt{\lambda_1}} X \mathbf{u}_1$$

satisfies the foregoing equations. (The eigenvalue is necessarily positive, since if zero there are no associated eigenvectors.) Similarly,

$$\mathbf{v}_k = \frac{1}{\sqrt{\lambda_k}} X \mathbf{u}_k$$

and

$$\mathbf{u}_k = \frac{1}{\sqrt{\lambda_k}} X' \mathbf{v}_k$$

for the kth largest eigenvalues and eigenvectors (or principal axes). Thus successive eigenvalues in both spaces are the same, and there is a simple linear transformation which maps the optimal axes in one space into those in the other.

The variance of the projections on a given axis in $\mathbb{R}^m$ is given by $(X\mathbf{u})'(X\mathbf{u})$, which by the eigenvector equation, is seen to equal λ.

In some software packages, the eigenvectors are rescaled so that $\sqrt{\lambda}\mathbf{u}$ and $\sqrt{\lambda}\mathbf{v}$ are used instead of $\mathbf{u}$ and $\mathbf{v}$. In this case, the *factor* $\sqrt{\lambda}\mathbf{u}$ gives the new, rescaled projections of the points in the space $\mathbb{R}^n$ (i.e. $\sqrt{\lambda}\mathbf{u} = X'\mathbf{v}$).

The coordinates of the new axes can be written in terms of the old coordinate system. Since

$$\mathbf{u} = \frac{1}{\sqrt{\lambda}} X' \mathbf{v}$$

each coordinate of the new vector $\mathbf{u}$ is defined as a linear combination of the initially-given vectors:

$$u_j = \sum_{i=1}^{n} \frac{1}{\sqrt{\lambda}} v_i x_{ij} = \sum_{i=1}^{n} c_i x_{ij}$$

(where $i \leq j \leq m$ and x_{ij} is the $(i,j)^{th}$ element of matrix X). Thus the jth coordinate of the new vector is a *synthetic* value formed from the jth coordinates of the given vectors (i.e. x_{ij} for all $1 \leq i \leq n$).

Since PCA in $\mathbb{R}^n$ and in $\mathbb{R}^m$ lead respectively to the finding of n and of m eigenvalues, and since in addition it has been seen that these eigenvalues are identical, it follows that the number of *non-zero eigenvalues* obtained in either space is less than or equal to $\min(n, m)$.

It has been seen that the eigenvectors associated with the p largest eigenvalues yield the best-fitting p-dimensional subspace of $\mathbb{R}^m$. A measure of the approximation is the percentage of variance explained by the subspace

$$\sum_{k \leq p} \lambda_k / \sum_{k=1}^{n} \lambda_k$$

expressed as a percentage.

An alternative view of PCA as an approximation to the given points is as follows. Taking the relationship between the unitary vectors in $\mathbb{R}^n$ and $\mathbb{R}^m$, $X\mathbf{u}_k = \sqrt{\lambda_k} \mathbf{v}_k$, postmultiplying by $\mathbf{u}'_k$, and summing gives

$$X \sum_{k=1}^{n} \mathbf{u}_k \mathbf{u}'_k = \sum_{k=1}^{n} \sqrt{\lambda_k} \mathbf{v}_k \mathbf{u}'_k.$$

The summation on the left hand side gives the identity matrix (this follows from the orthogonality of eigenvectors, so that the off-diagonal elements are zero; and the fact that the eigenvectors are of unit norm, so that the diagonal elements equal one), and so

$$X = \sum_{k=1}^{n} \sqrt{\lambda_k} \mathbf{v}_k \mathbf{u}'_k.$$

This series expansion of the given matrix, X, is termed the Karhunen–Loève expansion. The p best eigenvectors therefore approximate the original matrix by

2.2. MATHEMATICAL DESCRIPTION

$$X = \sum_{k=1}^{p} \sqrt{\lambda_k}\, \mathbf{v}_k \mathbf{u}'_k$$

and if p is very much less than n, and X approximately equal to X, an appreciable economy in description has been obtained.

2.2.5 Practical Aspects

Since the variables or attributes under analysis are often very different (some will "shout louder" than others), it is usual to standardize each variable in the following way. If r_{ij} are the original measurements, then the matrix X of (i,j)-value

$$x_{ij} = \frac{r_{ij} - \bar{r}_j}{s_j \sqrt{n}}$$

where

$$\bar{r}_j = \frac{1}{n} \sum_{i=1}^{n} r_{ij}$$

and

$$s_j^2 = \frac{1}{n} \sum_{i=1}^{n} (r_{ij} - \bar{r}_j)^2$$

is submitted to the PCA (cf. Chapter 1, where standardization was discussed: the multiplicative constant, $1/\sqrt{n}$, in the definition of x_{ij} above is used so that we may conveniently define correlations). The matrix to be diagonalized, $X'X$, is then of $(j,k)^{th}$ term:

$$\rho_{jk} = \sum_{i=1}^{n} x_{ij} x_{ik} = \frac{1}{n} \sum_{i=1}^{n} (r_{ij} - \bar{r}_j)(r_{ik} - \bar{r}_k) / s_j s_k$$

which is the correlation coefficient between variables j and k.

Using the definitions of x_{ij} and s_j above, the distance between variables j and k is

$$d^2(j,k) = \sum_{i=1}^{n} (x_{ij} - x_{ik})^2 = \sum_{i=1}^{n} x_{ij}^2 + \sum_{i=1}^{n} x_{ik}^2 - 2 \sum_{i=1}^{n} x_{ij} x_{ik}$$

and, substituting, the first two terms both yield 1, giving

$$d^2(j,k) = 2(1 - \rho_{jk}).$$

Thus the distance between variables is directly proportional to the correlation between them.

The distance between row vectors is

$$d^2(i,h) = \sum_j (x_{ij} - x_{hj})^2 = \sum_j (\frac{r_{ij} - r_{hj}}{\sqrt{n}s_j})^2 = (\mathbf{r}_i - \mathbf{r}_h)'M(\mathbf{r}_i - \mathbf{r}_h)$$

where $\mathbf{r}_i$ and $\mathbf{r}_h$ are column vectors (of dimensions $m \times 1$) and M is the $m \times m$ diagonal matrix of jth element $1/ns_j^2$. Therefore d is a Euclidean distance associated with matrix M. Note that the row points are now centred but the column points are not: therefore the latter may well appear in one quadrant on output listings.

Analysis of the matrix of $(j,k)^{th}$ term ρ_{jk} as defined above is PCA on a *correlation* matrix. Inherent in this, as has been seen, is that the row vectors are centred and reduced.

If, instead, centring was acceptable but not the rescaling of the variance, we would be analysing the matrix of $(j,k)^{th}$ term

$$c_{jk} = \frac{1}{n} \sum_{i=1}^n (r_{ij} - \bar{r}_j)(r_{ik} - \bar{r}_k).$$

In this case we have PCA of the *variance–covariance* matrix.

The following should be noted.

- We may speak about carrying out a PCA on the correlation matrix, on a variance–covariance matrix, or on the "sums of squares and cross–products" matrix. These relate to intermediate matrices which are most often determined by the PCA program. The user should however note the effect of transformations on his/her data (standardization, centring) which are inherent in these different options.

- Rarely is it recommendable to carry out a PCA on the "sums of squares and cross–products" matrix; instead some transformation of the original data is usually necessary. In the absence of justification for treating the data otherwise, the most recommendable strategy is to use the option of PCA on a correlation matrix.

- All the quantities defined above (standard deviation, variances and covariances, etc.) have been in *population* rather than in *sample* terms. That is to say, the data under examination is taken as all we can immediately study, rather than being representative of a greater underlying distribution of points. Not all packages and program libraries share this viewpoint, and hence discrepancies may be detected between results obtained in practice from different sources.

2.2.6 Iterative Solution of Eigenvalue Equations

A simple iterative scheme for solving the eigenvector equation $Au = \lambda u$ is as follows.

Choose some trial vector, t_0 : e.g. $(1, 1, \ldots, 1)$. Then define $t_1, t_2, \ldots$:

$$At_0 = x_0 \qquad t_1 = x_0/\sqrt{x_0'x_0}$$
$$At_1 = x_1 \qquad t_2 = x_1/\sqrt{x_1'x_1}$$
$$At_2 = x_2 \qquad t_3 = \ldots$$

We halt when there is sufficient convergence: $|t_n - t_{n+1}| \leq \epsilon$, for some small, real ϵ ; or when the number of iterations exceeds some maximum (e.g. 15). In the latter case, another choice of t_0 will be tried.

If there is convergence, $t_n = t_{n+1}$, we have the following:

$$At_n = x_n$$
$$t_{n+1} = x_n/\sqrt{x_n'x_n}.$$

Substituting for x_n in the first of these two equations gives

$$At_n = \sqrt{x_n'x_n}\ t_{n+1}.$$

Hence, since $t_n = t_{n+1}$, t_n is none other than the eigenvector sought; and the associated eigenvalue is $\sqrt{x_n'x_n}$.

The second eigenvector and associated eigenvalue may be found by carrying out a similar iterative algorithm on a matrix where the effects of u_1 and λ_1 have been *partialled out*: $A_{(2)} = A - \lambda_1 u_1 u_1'$. Let us prove that $A_{(2)}$ removes the effects due to the first eigenvector and eigenvalue. We have $Au = \lambda u$. Therefore $Auu' = \lambda uu'$; or equivalently, $Au_k u_k' = \lambda_k u_k u_k'$ for each eigenvalue. Summing over k gives:

$$A \sum_k u_k u_k' = \sum_k \lambda_k u_k u_k'.$$

The summed term on the left hand side equals the identity matrix (this has already been seen in Section 2.2.4 above). Therefore we have

$$A = \lambda_1 u_1 u_1' + \lambda_2 u_2 u_2' + \ldots$$

From this *spectral decomposition* of matrix A, we may successively remove the effects of the eigenvectors and eigenvalues as they are obtained.

Many other algorithms are available for solving eigen–equations: Chambers (1977) may be consulted for algorithms which are more likely to be implemented in the major commercial packages or subroutine libraries; and Smith et al. (1976) contains many Fortran implementations.

2.3 Examples and Bibliography

2.3.1 General Remarks

Among the objectives of PCA are the following:

1. dimensionality reduction;

2. the determining of linear combinations of variables;

3. feature selection: the choosing of the most useful variables;

4. visualization of multidimensional data;

5. identification of underlying variables;

6. identification of groups of objects or of outliers.

The tasks required of the analyst to carry these out are as follows:

1. In the case of a table of dimensions $n \times m$, each of the n rows or objects can be regarded as an m–dimensional vector. Finding a set of $m' < m$ principal axes allows the objects to be adequately characterised on a smaller number of (artificial) variables. This is advantageous as a prelude to further analysis as the $m - m'$ dimensions may often be ignored as constituting noise; and, secondly, for storage economy (sufficient information from the initial table is now represented in a table with $m' < m$ columns). Reduction of dimensionality is practicable if the first m' new axes account for approximately 75 % or more of the variance. There is no set threshold, — the analyst must judge. The cumulative percentage of variance explained by the principal axes is consulted in order to make this choice.

2. If the eigenvalue is zero, the variance of projections on the associated eigenvector is zero. Hence the eigenvector is reduced to a point. If this point is additionally the origin (i.e. the data are centred), then we have $X\mathbf{u} = \lambda\mathbf{u} = \mathbf{0}$. I.e. $\sum_j u_j \mathbf{x}_j = 0$, where $\mathbf{x}_j$ is the jth column vector of X. This allows linear combinations between the variables to be found. In fact, we can go a good deal further: by analysing second–order variables, defined from the given variables, quadratic dependencies can be staightforwardly sought. This means, for example, that in analysing three variables, y_1, y_2, and y_3, we would also input the variables y_1^2, y_2^2, y_3^2, $y_1 y_2$, $y_1 y_3$, and $y_2 y_3$. If the linear combination

$$y_1 = c_1 y_2^2 + c_2 y_1 y_2$$

exists, then we would find it. Similarly we could feed in the logarithms or other functions of variables.

3. In feature selection we want to simplify the task of characterising each object by a set of attributes. Linear combinations among attributes must be found; highly correlated attributes (i.e. closely located attributes in the new space: cf. section 2.2.5) allow some attributes to be removed from consideration; and the proximity of attributes to the new axes indicate the more relevant and important attributes.

4. In order to provide a convenient representation of multidimensional data, planar plots are necessary. An important consideration is the adequacy of the planar representation: the percentage variance explained by the pair of axes defining the plane must be looked at here.

5. PCA is often motivated by the search for latent variables. Often it is relatively easy to label the highest or second highest components, but it becomes increasingly difficult as less relevant axes are examined. The objects with the highest loadings or projections on the axes (i.e. those which are placed towards the extremities of the axes) are usually worth examining: the axis may be characterisable as a spectrum running from a small number of objects with high positive loadings to those with high negative loadings.

6. A visual inspection of a planar plot indicates which objects are grouped together, thus indicating that they belong to the same family or result from the same process. Anomalous objects can also be detected, and in some cases it might be of interest to redo the analysis with these excluded because of the perturbation they introduce.

2.3.2 Artificial Data

Data of the following characteristics was generated in order to look at the use of PCA for ascertaining quadratic dependencies. Thirty objects were used, and in total 5 variables.

$y_{1j} = -1.4, -1.3, \ldots, 1.5$

$y_{2j} = 2.0 - y_{1j}^2$

$y_{3j} = y_{1j}^2$

$y_{4j} = y_{2j}^2$

$y_{5j} = y_{1j} y_{2j}$

The output obtained follows. We could plot the row–points or column–points in the principal plane, using the coordinates found. The fifth eigenvalue is zero; hence the variance of the associated principal component is zero. Since we know in addition that the eigenvectors are centred, we therefore have the equation:

$$0.7071 y_2 + 0.7071 y_3 = 0.0$$

Note again that variables y_2 and y_3 have been redefined so that each value is centred (see section 2.2.5 above regarding PCA of a covariance matrix).

```
COVARIANCE MATRIX FOLLOWS.

  22.4750
  -2.2475    13.6498
   2.2475   -13.6498    13.6498
  -2.9262    28.0250   -28.0250    62.2917
  14.5189     0.5619    -0.5619     0.7316    17.3709


EIGENVALUES FOLLOW.

Eigenvalues      As Percentages    Cumul. Percentages
-----------      --------------    ------------------
  88.3852           68.2842              68.2842
  34.5579           26.6985              94.9828
   5.2437            4.0512              99.0339
   1.2505            0.9661             100.0000
   0.0000            0.0000             100.0000


EIGENVECTORS FOLLOW.

VBLE.    EV-1      EV-2      EV-3      EV-4      EV-5
-----   ------    ------    ------    ------    ------
  1    -0.0630    0.7617    0.6242   -0.1620    0.0000
  2     0.3857    0.0067   -0.1198   -0.5803    0.7071
  3    -0.3857   -0.0067    0.1198    0.5803    0.7071
  4     0.8357    0.0499    0.1593    0.5232    0.0000
  5     0.0018    0.6460   -0.7458    0.1627    0.0000


PROJECTIONS OF ROW-POINTS FOLLOW.

OBJECT  PROJ-1    PROJ-2    PROJ-3    PROJ-4    PROJ-5
------  ------    ------    ------    ------    ------
   1   -2.5222   -1.2489   -0.9036    0.5779    0.0000
   2   -2.2418   -1.3886   -0.6320    0.2413    0.0000
   3   -1.8740   -1.4720   -0.3942    0.0050    0.0000
   4   -1.4437   -1.5046   -0.1905   -0.1478    0.0000
   5   -0.9741   -1.4915   -0.0209   -0.2324    0.0000
```

2.3. EXAMPLES AND BIBLIOGRAPHY

```
 6   -0.4862 -1.4379  0.1153 -0.2630  0.0000
 7    0.0009 -1.3488  0.2187 -0.2525  0.0000
 8    0.4702 -1.2292  0.2906 -0.2125  0.0000
 9    0.9065 -1.0837  0.3327 -0.1535  0.0000
10    1.2969 -0.9171  0.3469 -0.0845  0.0000
11    1.6303 -0.7338  0.3355 -0.0136  0.0000
12    1.8977 -0.5383  0.3014  0.0528  0.0000
13    2.0920 -0.3349  0.2477  0.1093  0.0000
14    2.2083 -0.1278  0.1779  0.1516  0.0000
15    2.2434  0.0791  0.0958  0.1771  0.0000
16    2.1964  0.2817  0.0059  0.1840  0.0000
17    2.0683  0.4762 -0.0873  0.1720  0.0000
18    1.8620  0.6590 -0.1787  0.1420  0.0000
19    1.5826  0.8264 -0.2629  0.0962  0.0000
20    1.2371  0.9751 -0.3341  0.0381  0.0000
21    0.8345  1.1016 -0.3860 -0.0278  0.0000
22    0.3858  1.2027 -0.4121 -0.0955  0.0000
23   -0.0959  1.2755 -0.4055 -0.1578  0.0000
24   -0.5957  1.3168 -0.3587 -0.2062  0.0000
25   -1.0964  1.3238 -0.2641 -0.2312  0.0000
26   -1.5792  1.2938 -0.1135 -0.2216  0.0000
27   -2.0227  1.2242  0.1015 -0.1653  0.0000
28   -2.4042  1.1124  0.3898 -0.0489  0.0000
29   -2.6984  0.9561  0.7607  0.1424  0.0000
30   -2.8783  0.7530  1.2238  0.4245  0.0000


PROJECTIONS OF COLUMN-POINTS FOLLOW.

VBLE.  PROJ-1 PROJ-2 PROJ-3 PROJ-4 PROJ-5
------ ------ ------ ------ ------ ------
  1   -0.6739  2.7008  0.3289 -0.0203  0.0000
  2    4.1259  0.0236 -0.0632 -0.0726  0.0000
  3   -4.1259 -0.0236  0.0632  0.0726  0.0000
  4    8.9388  0.1768  0.0840  0.0654  0.0000
  5    0.0196  2.2906 -0.3930  0.0203  0.0000
```

2.3.3 Examples from Astronomy

PCA has been a fairly widely used technique in astronomy. The following list does not aim to be comprehensive, but indicates instead the types of problems to which PCA can be applied. It is also hoped that it may provide a convenient entry-point to literature on a topic of interest. References below are concerned with stellar parallaxes; a large number are concerned with the study of galaxies; and a large number relate also to spectral reduction.

1. A. Bijaoui, "Application astronomique de la compression de l'information", *Astronomy and Astrophysics*, **30**, 199–202, 1974.

2. A. Bijaoui, SAI Library, Algorithms for Image Processing, Nice Observatory, Nice, 1985.

   (A large range of subroutines for image processing, including the Karhunen–Loève expansion.)

3. P. Brosche, "The manifold of galaxies: Galaxies with known dynamical properties", *Astronomy and Astrophysics*, **23**, 259–268, 1973.

4. P. Brosche and F.T. Lentes, "The manifold of globular clusters", *Astronomy and Astrophysics*, **139**, 474–476, 1984.

5. V. Bujarrabal, J. Guibert and C. Balkowski, "Multidimensional statistical analysis of normal galaxies", *Astronomy and Astrophysics*, **104**, 1–9, 1981.

6. R. Buser, "A systematic investigation of multicolor photometric systems. I. The UBV, RGU and *uvby* systems.", *Astronomy and Astrophysics*, **62**, 411–424, 1978.

7. C.A. Christian and K.A. Janes, "Multivariate analysis of spectrophotometry". *Publications of the Astronomical Society of the Pacific*, **89**, 415–423, 1977.

8. C.A. Christian, "Identification of field stars contaminating the colour–magnitude diagram of the open cluster Be 21", *The Astrophysical Journal Supplement Series*, **49**, 555–592, 1982.

9. T.J. Deeming, "Stellar spectral classification. I. Application of component analysis", *Monthly Notices of the Royal Astronomical Society*, **127**, 493–516, 1964.

   (An often referenced work.)

10. T.J. Deeming, "The analysis of linear correlation in astronomy", *Vistas in Astronomy*, **10**, 125, 1968.

    (For regression also.)

11. G. Efstathiou and S.M. Fall, "Multivariate analysis of elliptical galaxies", *Monthly Notices of the Royal Astronomical Society*, **206**, 453–464, 1984.

12. S.M. Faber, "Variations in spectral–energy distributions and absorption–line strengths among elliptical galaxies", *The Astrophysical Journal*, **179**, 731–754, 1973.

13. M. Fofi, C. Maceroni, M. Maravalle and P. Paolicchi, "Statistics of binary stars. I. Multivariate analysis of spectroscopic binaries", *Astronomy and Astrophysics*, **124**, 313–321, 1983.

    (PCA is used, together with a non–hierarchical clustering technique.)

2.3. EXAMPLES AND BIBLIOGRAPHY

14. M. Fracassini, L.E. Pasinetti, E. Antonello and G. Raffaelli, "Multivariate analysis of some ultrashort period Cepheids (USPC)", *Astronomy and Astrophysics*, **99**, 397–399, 1981.

15. M. Fracassini, G. Manzotti, L.E. Pasinetti, G. Raffaelli, E. Antonello and L. Pastori, "Application of multivariate analysis to the parameters of astrophysical objects", in *Statistical Methods in Astronomy*, European Space Agency Special Publication 201, 21–25, 1983.

16. P. Galeotti, "A statistical analysis of metallicity in spiral galaxies", *Astrophysics and Space Science*, **75**, 511–519, 1981.

17. A. Heck, "An application of multivariate statistical analysis to a photometric catalogue", *Astronomy and Astrophysics*, **47**, 129–135, 1976.

    (PCA is used, along with regression and discriminant analysis.)

18. A. Heck, D. Egret, Ph. Nobelis and J.C. Turlot, "Statistical confirmation of the UV spectral classification system based on IUE low–dispersion spectra", *Astrophysics and Space Science*, **120**, 223–237, 1986.

    (Many other articles by these authors, which also make use of PCA, are referenced in the above.)

19. S.J. Kerridge and A.R. Upgren, "The application of multivariate analysis to parallax solutions. II. Magnitudes and colours of comparison stars", *The Astronomical Journal*, **78**, 632–638, 1973.

    (See also Upgren and Kerridge, 1971, referenced below.)

20. J. Koorneef, "On the anomaly of the far UV extinction in the 30 Doradus region", *Astronomy and Astrophysics*, **64**, 179–193, 1978.

    (PCA is used for deriving a photometric index from 5–channel photometric data.)

21. M.J. Kurtz, "Automatic spectral classification", PhD Thesis, Dartmouth College, New Hampshire, 1982.

22. F.T. Lentes, "The manifold of spheroidal galaxies", *Statistical Methods in Astronomy*, European Space Agency Special Publication 201, 73–76, 1983.

23. D. Massa and C.F. Lillie, "Vector space methods of photometric analysis: applications to O stars and interstellar reddening", *The Astrophysical Journal*, **221**, 833–850, 1978.

24. D. Massa, "Vector space methods of photometric analysis. III. The two components of ultraviolet reddening", *The Astronomical Journal*, **85**, 1651–1662, 1980.

25. B. Nicolet, "Geneva photometric boxes. I. A topological approach of photometry and tests.", *Astronomy and Astrophysics*, **97**, 85–93, 1981.

    (PCA is used on colour indices.)

26. S. Okamura, K. Kodaira and M. Watanabe, "Digital surface photometry of galaxies toward a quantitative classification. III. A mean concentration index as a parameter representing the luminosity distribution", *The Astrophysical Journal*, **280**, 7–14, 1984.

27. S. Okamura, "Global structure of Virgo cluster galaxies", in O.-G. Richter and B. Binggeli (eds.), Proceedings of ESO Workshop on The Virgo Cluster of Galaxies, ESO Conference and Workshop Proceedings No. 20, 201–215, 1985.

28. D. Pelat, "A study of H I absorption using Karhunen–Loève series", *Astronomy and Astrophysics*, **40**, 285–290, 1975.

29. A. W. Strong, "Data analysis in gamma–ray astronomy: multivariate likelihood method for correlation studies", *Astronomy and Astrophysics*, **150**, 273–275, 1985.

    (The method presented is not linked to PCA, but in dealing with the eigenreduction of a correlation matrix it is clearly very closely related.)

30. B. Takase, K. Kodaira and S. Okamura, *An Atlas of Selected Galaxies*, University of Tokyo Press, VNU Science Press, 1984.

31. D.J. Tholen, "Asteroid taxonomy from cluster analysis of photometry", PhD Thesis, University of Arizona, 1984.

32. A.R. Upgren and S.J. Kerridge, "The application of multivariate analysis to parallax solutions. I. Choice of reference frames", *The Astronomical Journal*, **76**, 655–664, 1971.

    (See also Kerridge and Upgren, 1973, referenced above.)

33. J.P. Vader, "Multivariate analysis of elliptical galaxies in different environments", *The Astrophysical Journal*, **306**, 390–400, 1986.

    (The Virgo and Coma clusters are studied.)

34. C.A. Whitney, "Principal components analysis of spectral data. I. Methodology for spectral classification", *Astronomy and Astrophysics Supplement Series*, **51**, 443–461, 1983.

35. B.C. Whitmore, "An objective classification system for spiral galaxies. I. The two dominant dimensions", *The Astrophysical Journal*, **278**, 61–80, 1984.

2.3.4 General References

1. T.W. Anderson, *An Introduction to Multivariate Statistical Analysis*, Wiley, New York, 1984 (2nd ed.).

   (For inferential aspects relating to PCA.)

2. J.M. Chambers, *Computational Methods for Data Analysis*, Wiley, New York, 1977.

3. C. Chatfield and A.J. Collins, *Introduction to Multivariate Analysis*, Chapman and Hall, 1980.

   (A good introductory textbook.)

4. R. Gnanadesikan, *Methods for Statistical Data Analysis of Multivariate Observations*, Wiley, New York, 1977.

   (For details of PCA, clustering and discrimination.)

5. M. Kendall, *Multivariate Analysis*, Griffin, London, 1980 (2nd ed.).

   (Dated in relation to computing techniques, but exceptionally clear and concise in its treatment of many practical problems.)

6. L. Lebart, A. Morineau and K.M. Warwick, *Multivariate Descriptive Statistical Analysis*, Wiley, New York, 1984.

   (An excellent geometric treatment of PCA.)

7. F.H.C. Marriott, *The Interpretation of Multiple Observations*, Academic Press, New York, 1974.

   (A short, readable textbook.)

8. B.T. Smith *et al.*, *Matrix Eigensystem Routines — EISPACK Guide*, Lecture Notes in Computer Science 6, Springer Verlag, Berlin and New York, 1976.

2.4 Software and Sample Implementation

The following is a listing of the Principal Components Analysis program. Note that input and output is handled by subroutines. It may, in certain cases, be desirable to alter the output formats. Alternatively, there is a "switch" which allows no output to be produced and instead a driving routine can use information passed back in the arrays and vectors, as indicated.

No extra subroutines are required to run the program, except for a driving routine.

The following subroutines are called from the main PCA routine.

1. CORCOL determines correlations.
2. COVCOL determines covariances.
3. SCPCOL determines sums of squares and cross–products.
4. TRED2 reduces a symmetric matrix to tridiagonal form.
5. TQL2 derives eigenvalues and eigenvectors of a tridiagonal matrix.
6. OUTMAT outputs a matrix.
7. OUTHMT outputs a diagonal half–matrix.
8. OUTEVL outputs eigenvalues.
9. OUTEVC outputs eigenvectors.
10. OUTPRX outputs projections of row–points.
11. OUTPRY outputs projections of column–points.
12. PROJX determines projections of row–points.
13. PROJY determines projections of column–points.

2.4.1 Progam Listing

```
C+++++++++++++++++++++++++++++++++++++++++++++++++++++++++
C
C  Carry out a PRINCIPAL COMPONENTS ANALYSIS
C            (KARHUNEN-LOEVE EXPANSION).
C
C  To call: CALL PCA(N,M,DATA,METHOD,IPRINT,A1,W1,W2,A2,IERR)
C           where
C
C
C  N, M   : integer dimensions of ...
C  DATA   : input data.
C           On output, DATA contains in first 7 columns the
C           projections of the row-points on the first 7
C           principal components.
C  METHOD: analysis option.
C           = 1: on sums of squares & cross products matrix.
C           = 2: on covariance matrix.
C           = 3: on correlation matrix.
C  IPRINT: print options.
C           = 0: no printed output- arrays/vectors, only, contain
C                items calculated.
C           = 1: eigenvalues, only, output.
```

2.4. SOFTWARE AND SAMPLE IMPLEMENTATION

```
C            = 2: printed output, in addition, of correlation (or
C                 other) matrix, eigenvalues and eigenvectors.
C            = 3: full printing of items calculated.
C    A1    : correlation, covariance or sums of squares &
C            cross-products matrix, dimensions M * M.
C            On output, A1 contains in the first 7 columns the
C            projections of the column-points on the first 7
C            principal components.
C    W1,W2 : real vectors of dimension M (see called routines for
C            use).
C            On output, W1 contains the cumulative percentage
C            variances associated with the principal components.
C    A2    : real array of dimensions M * M (see routines for use).
C    IERR  : error indicator (normally zero).
C
C
C  Inputs here are N, M, DATA, METHOD, IPRINT (and IERR).
C  Output information is contained in DATA, A1, and W1.
C  All printed outputs are carried out in easily recognizable sub-
C  routines called from the first subroutine following.
C
C  If IERR > 0, then its value indicates the eigenvalue for which
C  no convergence was obtained.
C
C-----------------------------------------------------------------
       SUBROUTINE PCA(N,M,DATA,METHOD,IPRINT,A,W,FV1,Z,IERR)
       REAL    DATA(N,M), A(M,M), W(M), FV1(M), Z(M,M)
C
       IF (METHOD.EQ.1) GOTO 100
       IF (METHOD.EQ.2) GOTO 400
C      If method.eq.3 or otherwise ...
       GOTO 700
C
C         Form sums of squares and cross-products matrix.
C
  100  CONTINUE
       CALL SCPCOL(N,M,DATA,A)
C
       IF (IPRINT.GT.1) CALL OUTHMT(METHOD,M,A)
C
C         Now do the PCA.
C
       GOTO 1000
C
C         Form covariance matrix.
C
  400  CONTINUE
       CALL COVCOL(N,M,DATA,W,A)
C
```

CHAPTER 2. PRINCIPAL COMPONENTS ANALYSIS

```
              IF (IPRINT.GT.1) CALL OUTHMT(METHOD,M,A)

C
C             Now do the PCA.
C
              GOTO 1000
C
C             Construct correlation matrix.
C
     700      CONTINUE
              CALL CORCOL(N,M,DATA,W,FV1,A)
C
              IF (IPRINT.GT.1) CALL OUTHMT(METHOD,M,A)

C
C             Now do the PCA.
C
              GOTO 1000
C
C             Carry out eigenreduction.
C
    1000      M2 = M
              CALL TRED2(M,M2,A,W,FV1,Z)
              CALL TQL2(M,M2,W,FV1,Z,IERR)
              IF (IERR.NE.0) GOTO 9000
C
C             Output eigenvalues and eigenvectors.
C
              IF (IPRINT.GT.0) CALL OUTEVL(N,M,W)
              IF (IPRINT.GT.1) CALL OUTEVC(N,M,Z)
C
C             Determine projections and output them.
C
              CALL PROJX(N,M,DATA,Z,FV1)
              IF (IPRINT.EQ.3) CALL OUTPRX(N,M,DATA)
              CALL PROJY(M,W,A,Z,FV1)
              IF (IPRINT.EQ.3) CALL OUTPRY(M,A)
C
    9000      RETURN
              END
C++++++++++++++++++++++++++++++++++++++++++++++++++
C
C  Determine correlations of columns.
C  First determine the means of columns, storing in WORK1.
C
C--------------------------------------------------------
              SUBROUTINE CORCOL(N,M,DATA,WORK1,WORK2,OUT)
              DIMENSION      DATA(N,M), OUT(M,M), WORK1(M), WORK2(M)
              DATA           EPS/1.E-10/
```

2.4. SOFTWARE AND SAMPLE IMPLEMENTATION

```
C
          DO 30 J = 1, M
             WORK1(J) = 0.0
             DO 20 I = 1, N
                WORK1(J) = WORK1(J) + DATA(I,J)
   20        CONTINUE
             WORK1(J) = WORK1(J)/FLOAT(N)
   30     CONTINUE
C
C         Next det. the std. devns. of cols., storing in WORK2.
C
          DO 50 J = 1, M
             WORK2(J) = 0.0
             DO 40 I = 1, N
                WORK2(J) = WORK2(J) + (DATA(I,J)
      X                    -WORK1(J))*(DATA(I,J)-WORK1(J))
   40        CONTINUE
             WORK2(J) = WORK2(J)/FLOAT(N)
             WORK2(J) = SQRT(WORK2(J))
             IF (WORK2(J).LE.EPS) WORK2(J) = 1.0
   50     CONTINUE
C
C         Now centre and reduce the column points.
C
          DO 70 I = 1, N
             DO 60 J = 1, M
                DATA(I,J) = (DATA(I,J)
      X                    -WORK1(J))/(SQRT(FLOAT(N))*WORK2(J))
   60        CONTINUE
   70     CONTINUE
C
C         Finally calc. the cross product of the data matrix.
C
          DO 100 J1 = 1, M-1
             OUT(J1,J1) = 1.0
             DO 90 J2 = J1+1, M
                OUT(J1,J2) = 0.0
                DO 80 I = 1, N
                   OUT(J1,J2) = OUT(J1,J2) + DATA(I,J1)*DATA(I,J2)
   80           CONTINUE
                OUT(J2,J1) = OUT(J1,J2)
   90        CONTINUE
  100     CONTINUE
          OUT(M,M) = 1.0
C
          RETURN
          END
C+++++++++++++++++++++++++++++++++++++++++++++++++
C
```

```
C   Determine covariances of columns.
C   First determine the means of columns, storing in WORK.
C
C-----------------------------------------------------------
        SUBROUTINE COVCOL(N,M,DATA,WORK,OUT)
        DIMENSION       DATA(N,M), OUT(M,M), WORK(M)
C
        DO 30 J = 1, M
           WORK(J) = 0.0
           DO 20 I = 1, N
              WORK(J) = WORK(J) + DATA(I,J)
   20      CONTINUE
           WORK(J) = WORK(J)/FLOAT(N)
   30   CONTINUE
C
C       Now centre the column points.
C
        DO 50 I = 1, N
           DO 40 J = 1, M
              DATA(I,J) = DATA(I,J)-WORK(J)
   40      CONTINUE
   50   CONTINUE
C
C       Finally calculate the cross product matrix of the
C       redefined data matrix.
C
        DO 80 J1 = 1, M
           DO 70 J2 = J1, M
              OUT(J1,J2) = 0.0
              DO 60 I = 1, N
                 OUT(J1,J2) = OUT(J1,J2) + DATA(I,J1)*DATA(I,J2)
   60         CONTINUE
              OUT(J2,J1) = OUT(J1,J2)
   70      CONTINUE
   80   CONTINUE
C
        RETURN
        END
C++++++++++++++++++++++++++++++++++++++++++++++++++++++++++
C
C   Determine sums of squares and cross-products of columns.
C
C-----------------------------------------------------------
        SUBROUTINE SCPCOL(N,M,DATA,OUT)
        DIMENSION       DATA(N,M), OUT(M,M)
C
        DO 30 J1 = 1, M
           DO 20 J2 = J1, M
              OUT(J1,J2) = 0.0
```

2.4. SOFTWARE AND SAMPLE IMPLEMENTATION

```
              DO 10 I = 1, N
                 OUT(J1,J2) = OUT(J1,J2) + DATA(I,J1)*DATA(I,J2)
 10           CONTINUE
              OUT(J2,J1) = OUT(J1,J2)
 20        CONTINUE
 30     CONTINUE
C
        RETURN
        END
C+++++++++++++++++++++++++++++++++++++++++++++++++++++
C
C Reduce a real, symmetric matrix to a symmetric, tridiagonal
C matrix.
C
C To call:    CALL TRED2(NM,N,A,D,E,Z)    where
C
C NM = row dimension of A and Z;
C N  = order of matrix A (will always be <= NM);
C A  = symmetric matrix of order N to be reduced to tridiag. form;
C D  = vector of dim. N containing, on output, diagonal elts. of
C      tridiagonal matrix.
C E  = working vector of dim. at least N-1 to contain subdiagonal
C      elements.
C Z  = matrix of dims. NM by N containing, on output, orthogonal
C      transformation matrix producing the reduction.
C
C Normally a call to TQL2 will follow the call to TRED2 in order to
C produce all eigenvectors and eigenvalues of matrix A.
C
C Algorithm used: Martin et al., Num. Math. 11, 181-195, 1968.
C
C Reference: Smith et al., Matrix Eigensystem Routines - EISPACK
C Guide, Lecture Notes in Computer Science 6, Springer-Verlag,
C 1976, pp. 489-494.
C
C-------------------------------------------------------------------
        SUBROUTINE TRED2(NM,N,A,D,E,Z)
        REAL A(NM,N),D(N),E(N),Z(NM,N)
C
        DO 100 I = 1, N
           DO 100 J = 1, I
              Z(I,J) = A(I,J)
 100    CONTINUE
        IF (N.EQ.1) GOTO 320
        DO 300 II = 2, N
           I = N + 2 - II
           L = I - 1
           H = 0.0
           SCALE = 0.0
```

```
              IF (L.LT.2) GOTO 130
              DO 120 K = 1, L
                 SCALE = SCALE + ABS(Z(I,K))
 120          CONTINUE
              IF (SCALE.NE.0.0) GOTO 140
 130          E(I) = Z(I,L)
              GOTO 290
 140          DO 150 K = 1, L
                 Z(I,K) = Z(I,K)/SCALE
                 H = H + Z(I,K)*Z(I,K)
 150          CONTINUE
C
              F = Z(I,L)
              G = -SIGN(SQRT(H),F)
              E(I) = SCALE * G
              H = H - F * G
              Z(I,L) = F - G
              F = 0.0
C
              DO 240 J = 1, L
                 Z(J,I) = Z(I,J)/H
                 G = 0.0
C                Form element of A*U.
                 DO 180 K = 1, J
                    G = G + Z(J,K)*Z(I,K)
 180             CONTINUE
                 JP1 = J + 1
                 IF (L.LT.JP1) GOTO 220
                 DO 200 K = JP1, L
                    G = G + Z(K,J)*Z(I,K)
 200             CONTINUE
C                Form element of P where P = I - U U' / H.
 220             E(J) = G/H
                 F = F + E(J) * Z(I,J)
 240          CONTINUE
              HH = F/(H + H)
C             Form reduced A.
              DO 260 J = 1, L
                 F = Z(I,J)
                 G = E(J) - HH * F
                 E(J) = G
                 DO 250 K = 1, J
                    Z(J,K) = Z(J,K) - F*E(K) - G*Z(I,K)
 250             CONTINUE
 260          CONTINUE
 290          D(I) = H
 300       CONTINUE
 320       D(1) = 0.0
           E(1) = 0.0
```

2.4. SOFTWARE AND SAMPLE IMPLEMENTATION

```
C         Accumulation of transformation matrices.
          DO 500 I = 1, N
             L = I - 1
             IF (D(I).EQ.0.0) GOTO 380
             DO 360 J = 1, L
                G = 0.0
                DO 340 K = 1, L
                   G = G + Z(I,K) * Z(K,J)
340             CONTINUE
                DO 350 K = 1, L
                   Z(K,J) = Z(K,J) - G * Z(K,I)
350             CONTINUE
360          CONTINUE
380          D(I) = Z(I,I)
             Z(I,I) = 1.0
             IF (L.LT.1) GOTO 500
             DO 400 J = 1, L
                Z(I,J) = 0.0
                Z(J,I) = 0.0
400          CONTINUE
500       CONTINUE
C
          RETURN
          END
C++++++++++++++++++++++++++++++++++++++++++++++++
C
C Determine eigenvalues and eigenvectors of a symmetric,
C tridiagonal matrix.
C
C To call:    CALL TQL2(NM,N,D,E,Z,IERR)    where
C
C NM = row dimension of Z;
C N = order of matrix Z;
C D = vector of dim. N containing, on output, eigenvalues;
C E = working vector of dim. at least N-1;
C Z = matrix of dims. NM by N containing, on output, eigenvectors;
C IERR = error, normally 0, but 1 if no convergence.
C
C Normally the call to TQL2 will be preceded by a call to TRED2 in
C order to set up the tridiagonal matrix.
C
C Algorithm used: QL method of Bowdler et al., Num. Math. 11,
C 293-306, 1968.
C
C Reference: Smith et al., Matrix Eigensystem Routines - EISPACK
C Guide, Lecture Notes in Computer Science 6, Springer-Verlag,
C 1976, pp. 468-474.
C
C----------------------------------------------------------------
```

```
      SUBROUTINE TQL2(NM,N,D,E,Z,IERR)
      REAL    D(N), E(N), Z(NM,N)
      DATA    EPS/1.E-12/
C
      IERR = 0
      IF (N.EQ.1) GOTO 1001
      DO 100 I = 2, N
         E(I-1) = E(I)
  100 CONTINUE
      F = 0.0
      B = 0.0
      E(N) = 0.0
C
      DO 240 L = 1, N
         J = 0
         H = EPS * (ABS(D(L)) + ABS(E(L)))
         IF (B.LT.H) B = H
C        Look for small sub-diagonal element.
         DO 110 M = L, N
            IF (ABS(E(M)).LE.B) GOTO 120
C           E(N) is always 0, so there is no exit through
C           the bottom of the loop.
  110    CONTINUE
  120    IF (M.EQ.L) GOTO 220
  130    IF (J.EQ.30) GOTO 1000
         J = J + 1
C        Form shift.
         L1 = L + 1
         G = D(L)
         P = (D(L1)-G)/(2.0*E(L))
         R = SQRT(P*P+1.0)
         D(L) = E(L)/(P+SIGN(R,P))
         H = G-D(L)
C
         DO 140 I = L1, N
            D(I) = D(I) - H
  140    CONTINUE
C
         F = F + H
C        QL transformation.
         P = D(M)
         C = 1.0
         S = 0.0
         MML = M - L
C
         DO 200 II = 1, MML
            I = M - II
            G = C * E(I)
            H = C * P
```

2.4. SOFTWARE AND SAMPLE IMPLEMENTATION

```
              IF (ABS(P).LT.ABS(E(I))) GOTO 150
              C = E(I)/P
              R = SQRT(C*C+1.0)
              E(I+1) = S * P * R
              S = C/R
              C = 1.0/R
              GOTO 160
  150         C = P/E(I)
              R = SQRT(C*C+1.0)
              E(I+1) = S * E(I) * R
              S = 1.0/R
              C = C * S
  160         P = C * D(I) - S * G
              D(I+1) = H + S * (C * G + S * D(I))
C             Form vector.
              DO 180 K = 1, N
                 H = Z(K,I+1)
                 Z(K,I+1) = S * Z(K,I) + C * H
                 Z(K,I) = C * Z(K,I) - S * H
  180         CONTINUE
  200      CONTINUE
           E(L) = S * P
           D(L) = C * P
           IF (ABS(E(L)).GT.B) GOTO 130
  220      D(L) = D(L) + F
  240   CONTINUE
C
C       Order eigenvectors and eigenvalues.

        DO 300 II = 2, N
           I = II - 1
           K = I
           P = D(I)
           DO 260 J = II, N
              IF (D(J).GE.P) GOTO 260
              K = J
              P = D(J)
  260      CONTINUE
           IF (K.EQ.I) GOTO 300
           D(K) = D(I)
           D(I) = P
           DO 280 J = 1, N
              P = Z(J,I)
              Z(J,I) = Z(J,K)
              Z(J,K) = P
  280      CONTINUE
  300   CONTINUE
C
        GOTO 1001
C       Set error - no convergence after 30 iterns.
```

```
      1000    IERR = L
      1001    RETURN
              END
C++++++++++++++++++++++++++++++++++++++++++++++++++
C
C  Output array.
C
C-----------------------------------------------------------
              SUBROUTINE OUTMAT(N,M,ARRAY)
              DIMENSION ARRAY(N,M)
C
              DO 100 K1 = 1, N
                  WRITE (6,1000) (ARRAY(K1,K2),K2=1,M)
       100    CONTINUE
C
      1000    FORMAT(10(2X,F8.4))
              RETURN
              END
C++++++++++++++++++++++++++++++++++++++++++++++++++
C
C  Output half of (symmetric) array.
C
C-----------------------------------------------------------
              SUBROUTINE OUTHMT(ITYPE,NDIM,ARRAY)
              DIMENSION ARRAY(NDIM,NDIM)
C
              IF (ITYPE.EQ.1) WRITE (6,1000)
              IF (ITYPE.EQ.2) WRITE (6,2000)
              IF (ITYPE.EQ.3) WRITE (6,3000)
C
              DO 100 K1 = 1, NDIM
                  WRITE (6,4000) (ARRAY(K1,K2),K2=1,K1)
       100    CONTINUE
C
      1000    FORMAT
           X  (1H0,'SUMS OF SQUARES & CROSS-PRODUCTS MATRIX FOLLOWS.',/)
      2000    FORMAT(1H0,'COVARIANCE MATRIX FOLLOWS.',/)
      3000    FORMAT(1H0,'CORRELATION MATRIX FOLLOWS.',/)
      4000    FORMAT(8(2X,F8.4))
              RETURN
              END
C++++++++++++++++++++++++++++++++++++++++++++++++++
C
C  Output eigenvalues in order of decreasing value.
C
C-----------------------------------------------------------
              SUBROUTINE OUTEVL(N,NVALS,VALS)
              DIMENSION      VALS(NVALS)
C
```

2.4. SOFTWARE AND SAMPLE IMPLEMENTATION

```
              TOT = 0.0
              DO 100 K = 1, NVALS
                 TOT = TOT + VALS(K)
       100    CONTINUE
C
              WRITE (6,1000)
              CUM = 0.0
              K = NVALS + 1
C
              M = NVALS
C
C      (We only want Min(nrows,ncols) eigenvalues output:)
              M = MIN0(N,NVALS)
C
              WRITE (6,1010)
              WRITE (6,1020)
       200    CONTINUE
              K = K - 1
              CUM = CUM + VALS(K)
              VPC = VALS(K) * 100.0 / TOT
              VCPC = CUM * 100.0 / TOT
              WRITE (6,1030) VALS(K),VPC,VCPC
              VALS(K) = VCPC
              IF (K.GT.NVALS-M+1) GOTO 200
C
              RETURN
      1000    FORMAT(' EIGENVALUES FOLLOW.')
      1010    FORMAT
           X(' Eigenvalues       As Percentages    Cumul. Percentages')
      1020    FORMAT
           X(' -----------       --------------    ------------------')
      1030    FORMAT(F13.4,7X,F10.4,10X,F10.4)
              END
C++++++++++++++++++++++++++++++++++++++++++++++++++++++++
C
C        Output FIRST SEVEN eigenvectors associated with
C        eigenvalues in descending order.
C
C-----------------------------------------------------------
              SUBROUTINE OUTEVC(N,NDIM,VECS)
              DIMENSION     VECS(NDIM,NDIM)
C
              NUM = MIN0(N,NDIM,7)
C
              WRITE (6,1000)
              WRITE (6,1010)
              WRITE (6,1020)
              DO 100 K1 = 1, NDIM
              WRITE (6,1030) K1,(VECS(K1,NDIM-K2+1),K2=1,NUM)
       100    CONTINUE
C
```

```
      RETURN
1000  FORMAT(1H0,'EIGENVECTORS FOLLOW.',/)
1010  FORMAT
     X ('  VBLE.    EV-1    EV-2    EV-3    EV-4    EV-5    EV-6
     X    EV-7')
1020  FORMAT
     X ('  ------  ------  ------  ------  ------  ------  ------
     X------')
1030  FORMAT(I5,2X,7F8.4)
      END
C++++++++++++++++++++++++++++++++++++++++++++++++++++++++++++
C
C  Output projections of row-points on first 7 principal components.
C
C-------------------------------------------------------------
      SUBROUTINE OUTPRX(N,M,PRJN)
      REAL    PRJN(N,M)
C
      NUM = MIN0(M,7)
      WRITE (6,1000)
      WRITE (6,1010)
      WRITE (6,1020)
      DO 100 K = 1, N
          WRITE (6,1030) K,(PRJN(K,J),J=1,NUM)
  100 CONTINUE
C
1000  FORMAT(1H0,'PROJECTIONS OF ROW-POINTS FOLLOW.',/)
1010  FORMAT
     X ('  OBJECT  PROJ-1  PROJ-2  PROJ-3  PROJ-4  PROJ-5  PROJ-6
     X  PROJ-7')
1020  FORMAT
     X ('  ------  ------  ------  ------  ------  ------  ------
     X  ------')
1030  FORMAT(I5,2X,7F8.4)
      RETURN
      END
C++++++++++++++++++++++++++++++++++++++++++++++++++++++++++++
C
C  Output projections of columns on first 7 principal components.
C
C-------------------------------------------------------------
      SUBROUTINE OUTPRY(M,PRJNS)
      REAL    PRJNS(M,M)
C
      NUM = MIN0(M,7)
      WRITE (6,1000)
      WRITE (6,1010)
      WRITE (6,1020)
```

2.4. SOFTWARE AND SAMPLE IMPLEMENTATION

```
            DO 100 K = 1, M
               WRITE (6,1030) K,(PRJNS(K,J),J=1,NUM)
    100     CONTINUE
C
   1000     FORMAT(1H0,'PROJECTIONS OF COLUMN-POINTS FOLLOW.',/)
   1010     FORMAT
        X   ('  VBLE.   PROJ-1  PROJ-2  PROJ-3  PROJ-4  PROJ-5  PROJ-6
        X    PROJ-7')
   1020     FORMAT
        X   ('  ------  ------  ------  ------  ------  ------  ------
        X    ------')
   1030     FORMAT(I5,2X,7F8.4)
            RETURN
            END
C++++++++++++++++++++++++++++++++++++++++++++++++++++++++++++
C
C   Form projections of row-points on first 7 principal components.
C
C-------------------------------------------------------------
            SUBROUTINE PROJX(N,M,DATA,EVEC,VEC)
            REAL    DATA(N,M), EVEC(M,M), VEC(M)
C
            NUM = MIN0(M,7)
            DO 300 K = 1, N
               DO 50 L = 1, M
                  VEC(L) = DATA(K,L)
     50        CONTINUE
               DO 200 I = 1, NUM
                  DATA(K,I) = 0.0
                  DO 100 J = 1, M
                     DATA(K,I) = DATA(K,I) + VEC(J) *
        X                                   EVEC(J,M-I+1)
    100           CONTINUE
    200        CONTINUE
    300     CONTINUE
C
            RETURN
            END
C++++++++++++++++++++++++++++++++++++++++++++++++++++++++++++
C
C   Determine projections of column-points on 7 prin. components.
C
C-------------------------------------------------------------
            SUBROUTINE PROJY(M,EVALS,A,Z,VEC)
            REAL    EVALS(M), A(M,M), Z(M,M), VEC(M)
C
            NUM = MIN0(M,7)
            DO 300 J1 = 1, M
               DO 50 L = 1, M
```

```
                  VEC(L) = A(J1,L)
     50      CONTINUE
             DO 200 J2 = 1, NUM
                  A(J1,J2) = 0.0
                  DO 100 J3 = 1, M
                       A(J1,J2) = A(J1,J2) + VEC(J3)*Z(J3,M-J2+1)
    100           CONTINUE
                  IF (EVALS(M-J2+1).GT.0.00005) A(J1,J2) =
       X                     A(J1,J2)/SQRT(EVALS(M-J2+1))
                  IF (EVALS(M-J2+1).LE.0.00005) A(J1,J2) = 0.0
    200      CONTINUE
    300 CONTINUE
C
        RETURN
        END
```

2.4. SOFTWARE AND SAMPLE IMPLEMENTATION

2.4.2 Input Data

The following is the input data set used (Adorf, 1986). It represents a set of representative spectral intensity values versus wavelength for 18 main sequence stars. The reference spectra are of type O, B3, B5, B8, A0, A5, F0, F5, G0, G5, K0, K3, K5, K7, M0, M2, M4, and M5. More intensity values were initially present for each star, but in order to arrive at features of relevance the values at the beginning, end and middle of the wavelength range 350–530 nm were taken. The subsets of the original spectra, thus defined, encompassed the essential characteristics of downward sloping spectra being associated with O and B stars, and generally upward sloping spectra associated with K and M stars. The data used here have in total 16 intensity values, for each star.

```
INPUT DATA SET: STARS

Seq.no.    Col.1         Col.2         Col.3         Col.4

   1       3.00000       3.00000       3.00000       3.00000
   2       3.50000       3.50000       4.00000       4.00000
   3       4.00000       4.00000       4.50000       4.50000
   4       5.00000       5.00000       5.00000       5.50000
   5       6.00000       6.00000       6.00000       6.00000
   6      11.0000       11.0000       11.0000       11.0000
   7      20.0000       20.0000       20.0000       20.0000
   8      30.0000       30.0000       30.0000       30.0000
   9      30.0000       33.4000       36.8000       40.0000
  10      42.0000       44.0000       46.0000       48.0000
  11      60.0000       61.7000       63.5000       65.5000
  12      70.0000       70.1000       70.2000       70.3000
  13      78.0000       77.6000       77.2000       76.8000
  14      98.9000       97.8000       96.7000       95.5000
  15     160.000       157.000       155.000       152.000
  16     272.000       266.000       260.000       254.000
  17     382.000       372.000       362.000       352.000
  18     770.000       740.000       710.000       680.000

Seq.no.    Col.5         Col.6         Col.7         Col.8

   1       3.00000       3.00000      35.0000       45.0000
   2       4.50000       4.50000      46.0000       59.0000
   3       5.00000       5.00000      48.0000       60.0000
   4       5.50000       5.50000      46.0000       63.0000
```

Seq.no.				
5	6.50000	6.50000	51.0000	69.0000
6	11.0000	11.0000	64.0000	75.0000
7	20.0000	20.0000	76.0000	86.0000
8	30.1000	30.2000	84.0000	96.0000
9	43.0000	45.6000	100.000	106.000
10	50.0000	51.0000	109.000	111.000
11	67.3000	69.2000	122.000	124.000
12	70.4000	70.5000	137.000	132.000
13	76.4000	76.0000	167.000	159.000
14	94.3000	93.2000	183.000	172.000
15	149.000	147.000	186.000	175.000
16	248.000	242.000	192.000	182.000
17	343.000	333.000	205.000	192.000
18	650.000	618.000	226.000	207.000

Seq.no.	Col.9	Col.10	Col.11	Col.12
1	53.0000	55.0000	58.0000	113.000
2	63.0000	58.0000	58.0000	125.000
3	68.0000	65.0000	65.0000	123.000
4	70.0000	64.0000	63.0000	116.000
5	77.0000	70.0000	71.0000	120.000
6	81.0000	79.0000	79.0000	112.000
7	93.0000	92.0000	91.0000	104.000
8	98.0000	99.0000	96.0000	101.000
9	106.000	108.000	101.000	99.0000
10	110.000	110.000	103.000	95.5000
11	124.000	121.000	103.000	93.2000
12	134.000	128.000	101.000	91.7000
13	152.000	144.000	103.000	89.8000
14	162.000	152.000	102.000	87.5000
15	165.000	156.000	120.000	87.0000
16	170.000	159.000	131.000	88.0000
17	178.000	166.000	138.000	86.2000
18	195.000	180.000	160.000	82.9000

Seq.no.	Col.13	Col.14	Col.15	Col.16
1	113.000	86.0000	67.0000	90.0000
2	126.000	110.000	78.0000	97.0000
3	123.000	117.000	87.0000	108.000
4	119.000	115.000	97.0000	112.000
5	122.000	122.000	96.0000	123.000
6	114.000	113.000	98.0000	115.000
7	104.500	107.000	97.5000	104.000
8	102.000	99.0000	94.0000	99.0000
9	98.0000	99.0000	95.0000	95.0000
10	95.5000	95.0000	92.5000	92.0000
11	92.5000	92.2000	90.0000	90.8000

2.4. SOFTWARE AND SAMPLE IMPLEMENTATION

12	90.2000	88.8000	87.3000	85.8000
13	87.7000	85.7000	83.7000	81.8000
14	85.3000	83.3000	81.3000	79.3000
15	84.9000	82.8000	80.8000	79.0000
16	85.8000	83.7000	81.6000	79.6000
17	84.0000	82.0000	79.8000	77.5000
18	80.2000	77.7000	75.2000	72.7000

2.4.3 Sample Output

The following is the output which is produced. The "linearity" of the data may be noted. The eigenvalues (percentages of variance explained by axes) are followed by the definition of the eigenvectors (principal components) in the parameter–space. Then the projections of the objects (rows) and of the parameters (columns) on the new principal components in the respective spaces are listed.

EIGENVALUES FOLLOW.

Eigenvalues	As Percentages	Cumul. Percentages
12.7566	79.7285	79.7285
1.8521	11.5756	91.3041
1.1516	7.1975	98.5016
0.1762	1.1009	99.6026
0.0356	0.2226	99.8251
0.0225	0.1406	99.9658
0.0035	0.0219	99.9877
0.0007	0.0044	99.9921
0.0005	0.0033	99.9955
0.0004	0.0027	99.9982
0.0002	0.0015	99.9997
0.0001	0.0003	100.0000
0.0000	0.0000	100.0000
0.0000	0.0000	100.0000
0.0000	0.0000	100.0000
0.0000	0.0000	100.0000

EIGENVECTORS FOLLOW.

VBLE.	EV-1	EV-2	EV-3	EV-4	EV-5	EV-6	EV-7
1	0.2525	-0.3156	0.0281	0.0621	0.0475	0.1261	-0.0736
2	0.2534	-0.3105	0.0294	0.0623	0.0362	0.1088	-0.0493
3	0.2544	-0.3049	0.0305	0.0609	0.0215	0.0877	-0.0269
4	0.2555	-0.2990	0.0321	0.0613	0.0087	0.0700	0.0050
5	0.2565	-0.2927	0.0336	0.0594	-0.0065	0.0441	0.0374
6	0.2578	-0.2847	0.0350	0.0558	-0.0189	0.0144	0.0809
7	0.2678	0.1684	0.0717	-0.3855	0.0350	0.0712	0.1288
8	0.2665	0.1671	0.1065	-0.4025	0.0352	0.0189	0.4010
9	0.2663	0.1668	0.1267	-0.3752	0.1033	0.0802	-0.0830
10	0.2649	0.2000	0.1322	-0.2432	0.0217	-0.0583	-0.2050
11	0.2673	0.0493	0.2252	0.1838	-0.3041	-0.8467	0.0563
12	-0.2457	-0.3228	-0.1133	-0.3296	-0.2382	-0.1378	0.1235
13	-0.2489	-0.3182	-0.0784	-0.2706	-0.1544	-0.0606	0.5661
14	-0.2365	-0.2575	0.3337	-0.4175	-0.3199	-0.0074	-0.5839
15	-0.1354	0.0883	0.7990	0.2634	-0.2190	0.3249	0.2778
16	-0.2407	-0.2209	0.3530	-0.0914	0.8115	-0.3116	0.0123

PROJECTIONS OF ROW-POINTS FOLLOW.

OBJECT	PROJ-1	PROJ-2	PROJ-3	PROJ-4	PROJ-5	PROJ-6	PROJ-7
1	-0.6334	-0.1040	-0.7981	0.1659	0.0722	-0.0174	0.0010
2	-0.8520	-0.3140	-0.3998	-0.1283	-0.1015	0.0073	0.0096
3	-0.8994	-0.3214	-0.0674	-0.1203	-0.0423	-0.0136	-0.0257
4	-0.8988	-0.2452	0.1791	0.0087	0.0086	0.0907	0.0242
5	-0.9446	-0.3344	0.2805	-0.1295	0.0877	-0.0433	-0.0091
6	-0.7488	-0.1569	0.2835	0.0051	0.0547	-0.0080	0.0150
7	-0.4970	0.0263	0.2529	0.0898	-0.0213	-0.0022	-0.0215
8	-0.3373	0.1036	0.1385	0.1198	-0.0116	-0.0237	0.0167
9	-0.2170	0.1691	0.1792	0.1023	-0.0577	-0.0124	0.0002
10	-0.1077	0.2188	0.1092	0.1223	-0.0405	-0.0122	0.0004
11	0.0436	0.2570	0.0727	0.0736	0.0053	-0.0017	-0.0114
12	0.1606	0.3222	-0.0115	0.0371	0.0000	0.0322	-0.0155
13	0.3541	0.4244	-0.0727	-0.0862	0.0142	0.0312	-0.0009
14	0.4994	0.4539	-0.1191	-0.1304	0.0388	0.0573	-0.0051
15	0.6883	0.3365	-0.0761	-0.0903	0.0066	-0.0375	-0.0005
16	0.9433	0.0842	0.0061	-0.0685	-0.0123	-0.0485	0.0190
17	1.2569	-0.1027	0.0096	-0.0507	-0.0062	-0.0298	0.0150
18	2.1896	-0.8174	0.0333	0.0796	0.0053	0.0314	-0.0117

2.4. SOFTWARE AND SAMPLE IMPLEMENTATION

PROJECTIONS OF COLUMN-POINTS FOLLOW.

VBLE.	PROJ-1	PROJ-2	PROJ-3	PROJ-4	PROJ-5	PROJ-6	PROJ-7
1	0.3607	-0.0612	0.0033	0.0011	0.0002	0.0003	0.0000
2	0.3621	-0.0602	0.0034	0.0011	0.0001	0.0002	0.0000
3	0.3635	-0.0591	0.0035	0.0011	0.0001	0.0002	0.0000
4	0.3650	-0.0579	0.0037	0.0011	0.0000	0.0002	0.0000
5	0.3664	-0.0567	0.0039	0.0010	0.0000	0.0001	0.0000
6	0.3683	-0.0552	0.0041	0.0010	-0.0001	0.0000	0.0000
7	0.3826	0.0326	0.0083	-0.0068	0.0001	0.0002	0.0000
8	0.3807	0.0324	0.0124	-0.0071	0.0001	0.0000	0.0001
9	0.3804	0.0323	0.0147	-0.0066	0.0004	0.0002	0.0000
10	0.3784	0.0388	0.0153	-0.0043	0.0001	-0.0001	-0.0001
11	0.3819	0.0096	0.0261	0.0032	-0.0011	-0.0019	0.0000
12	-0.3511	-0.0626	-0.0131	-0.0058	-0.0008	-0.0003	0.0000
13	-0.3556	-0.0617	-0.0091	-0.0048	-0.0006	-0.0001	0.0002
14	-0.3378	-0.0499	0.0387	-0.0074	-0.0011	0.0000	-0.0002
15	-0.1934	0.0171	0.0927	0.0046	-0.0008	0.0007	0.0001
16	-0.3439	-0.0428	0.0410	-0.0016	0.0029	-0.0007	0.0000

Chapter 3

Cluster Analysis

3.1 The Problem

Automatic classification algorithms are used in widely different fields in order to provide a description or a reduction of data. A clustering algorithm is used to determine the inherent or natural groupings in the data, or provide a convenient summarization of the data into groups. Although the term "classification" is often applied both to this area and to Discriminant Analysis, this chapter will be solely concerned with unsupervised clustering, with no prior knowledge on the part of the analyst regarding group memberships.

As is the case with Principal Components Analysis, and with most other multivariate techniques, the objects to be classified have numerical measurements on a set of variables or attributes. Hence, the analysis is carried out on the rows of an array or matrix. If we have not a matrix of numerical values, to begin with, then it may be necessary to skilfully construct such a matrix. Chapter 1, in particular, can be referred to here for difficulties arising when missing data or qualitative attributes are present. The objects, or rows of the matrix, can be viewed as vectors in a multidimensional space (the dimensionality of this space being the number of variables or columns). A geometric framework of this type is not the only one which can be used to formulate clustering algorithms; it is the preferred one in this chapter because of its generality and flexibility.

Needless to say, a clustering analysis can just as easily be implemented on the columns of a data matrix. There is usually no direct relationship between the clustering of the rows and clustering of the columns, as was the case for the dual spaces in Principal Components Analysis. It may also be remarked that suitable alternative forms of storage of a rectangular array of values are not inconsistent with viewing the problem in geometric (and in matrix) terms: in the case of large sparse matrices, for instance, suitable storage schemes require consideration in practice.

Motivation for clustering may be categorized under the following headings, the first two of which will be of principal interest in this chapter:

1. Analysis of data: here, the given data is to be analyzed in order to reveal its fundamental features. The significant interrelationships present in the data are sought. This is the multivariate statistical use of clustering, and the validity problem (i.e. the validation of clusters of data items produced by an algorithm) is widely seen as a major current difficulty.

2. User convenience: a synoptic classification is to be obtained which will present a useful decomposition of the data. This may be a first step towards subsequent analysis where the emphasis is on heuristics for summarizing information. Appraisal of clustering algorithms used for this purpose can include algorithmic efficiency and ease of use.

3. Storage and retrieval: we are here concerned with improving access speeds by providing better routing strategies to stored information. The effectiveness of the clustering is measured by time and space efficiency, and by external criteria (related, for example, to the amount of relevant material retrieved).

4. Machine vision: the distinguishing of point patterns, or the processing of digital image data, is often assessed visually, and computational efficiency is highly important also.

Applications of clustering embrace many diverse fields (establishing taxonomies in biology and ecology; efficient retrieval algorithms in computer and information science; grouping of test subjects and of the test items in psychology and educational research; and so on). The range of algorithms which have been proposed (for the most part since the early 1960s with the advent of computing power on a wide scale) has been correspondingly large.

3.2 Mathematical Description

3.2.1 Introduction

Most published work in Cluster Analysis involves the use of either of two classes of clustering algorithm: hierarchical or non–hierarchical (often partitioning) algorithms. Hierarchical algorithms in particular have been dominant in the literature and so this chapter concentrates on these methods (Sections 3.2.2, 3.2.3). Each of the many clustering methods — and of the many hierarchical methods — which have been proposed over the last two decades have possibly advantageous properties. Many textbooks catalogue these methods, which makes the task facing the practitioner of choosing the right method an onerous one. We feel that it is more

3.2. MATHEMATICAL DESCRIPTION

helpful instead to study a method which is recommendable for general purpose applications. This we do with the minimum variance method.

Sections 3.2.4 and 3.2.5 will focus on Ward's minimum variance method. For most applications this method can be usefully employed for the summarization of data. Section 3.2.4 will attempt to justify this statement: it will informally describe the minimum variance agglomerative method, and will look at properties of this method which are of practical importance. Mathematical properties of the minimum variance method are then detailed in Section 3.2.5.

Finally, non–hierarchical routines have also been widely implemented. Section 3.2.6 looks at the minimal spanning tree (closely related to the single linkage hierarchical method). Finally, Section 3.2.7 briefly describes partitioning methods.

3.2.2 Hierarchical Methods

The single linkage hierarchical clustering approach outputs a set of clusters (to use graph theoretic terminology, a set of maximal connected subgraphs) at each level — or for each threshold value which produces a new partition. The following algorithm, in its general structure, is relevant for a wide range of hierarchical clustering methods which vary only in the update formula used in Step 2. These methods may, for example, define a criterion of compactness in Step 2 to be used instead of the connectivity criterion used here. Such hierarchical methods will be studied in Section 3.2.3, but the single linkage method with which we begin is one of the oldest and most widely used methods (its usage is usually traced to the early 1950s). An example is shown in Figure 3.1 — note that the dissimilarity coefficient is assumed to be symmetric, and so the clustering algorithm is implemented on half the dissimilarity matrix.

Single linkage hierarchical clustering

Input An $n(n-1)/2$ set of dissimilarities.

Step 1 Determine the smallest dissimilarity, d_{ik}.

Step 2 Agglomerate objects i and k: i.e. replace them with a new object, $i \cup k$; update dissimilarities such that, for all objects $j \neq i, k$:

$$d_{i \cup k, j} = \min \{d_{ij}, d_{kj}\}.$$

Delete dissimilarities d_{ij} and d_{kj}, for all j, as these are no longer used.

Step 3 While at least two objects remain, return to Step 1.

	1	2	3	4	5
1	0	4	9	5	8
2	4	0	6	3	6
3	9	6	0	6	3
4	5	3	6	0	5
5	8	6	3	5	0

Agglomerate 2 and 4 at dissimilarity 3.

	1	2U4	3	5
1	0	4	9	8
2U4	4	0	6	5
3	9	6	0	3
5	8	5	3	0

Agglomerate 3 and 5 at dissimilarity 3.

	1	2U4	3U5
1	0	4	8
2U4	4	0	5
3U5	8	5	0

Agglomerate 1 and 2U4 at dissimilarity 4.

	1U2U4	3U5
1U2U4	0	5
3U5	5	0

Finally agglomerate 1U2U4 and 3U5 at dissimilarity 5.

Resulting dendrogram: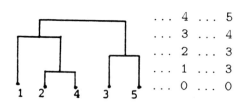

Ranks or levels.	Criterion values (or linkage weights).
... 4	... 5
... 3	... 4
... 2	... 3
... 1	... 3
... 0	... 0

Figure 3.1: Construction of a dendrogram by the single linkage method.

3.2. MATHEMATICAL DESCRIPTION

Equal dissimilarities may be treated in an arbitrary order. There are precisely $n-1$ agglomerations in Step 2 (allowing for arbitrary choices in Step 1 if there are identical dissimilarities). It may be convenient to index the clusters found in Step 2 by $n+1, n+2, \ldots, 2n-1$, or an alternative practice is to index cluster $i \cup k$ by the lower of the indices of i and k.

The title *single linkage* arises since, in Step 2, the interconnecting dissimilarity between two clusters ($i \cup k$ and j) or components is defined as the least interconnecting dissimilarity between a member of one and a member of the other. Other hierarchical clustering methods are characterized by other functions of the interconnecting linkage dissimilarities.

Since there are $n-1$ agglomerations, and hence iterations, and since Step 2 requires $< n$ operations, the algorithm for the single linkage hierarchic clustering given above is of time complexity $O(n^2)$.

Compared to other hierarchic clustering techniques, the single linkage method can give rise to a notable disadvantage for summarizing interrelationships. This is known as *chaining*. An example is to consider four subject-areas, which it will be supposed are characterized by certain attributes: computer science, statistics, probability, and measure theory. It is quite conceivable that "computer science" is connected to "statistics" at some threshold value, "statistics" to "probability", and "probability" to "measure theory", thereby giving rise to the fact that "computer science" and "measure theory" find themselves, undesirably, in the same cluster. This is due to the intermediaries "statistics" and "probability".

We will turn attention now to the general role played by a dendrogram, constructed by any criterion (Figure 3.2 illustrates differing possible representations).

About 75% of all published work on clustering has employed hierarchical algorithms (according to Blashfield and Aldenderfer, 1978): this figure for published work might or might not hold for practical clustering usage, but it is nonetheless revealing. Interpretation of the information contained in a dendrogram will often be of one or more of the following kinds:

– set inclusion relationships,

– partition of the object–sets, and

– significant clusters.

We will briefly examine what each of these entail.

Much early work on hierarchic clustering was in the field of biological taxonomy, from the 1950s and more so from the 1960s onwards. The central reference in this area, the first edition of which dates from the early 1960s, is Sneath and Sokal (1973). One major interpretation of hierarchies has been the evolution relationships between the organisms under study. It is hoped, in this context, that a dendrogram provides a sufficiently accurate model of underlying evolutionary pro-

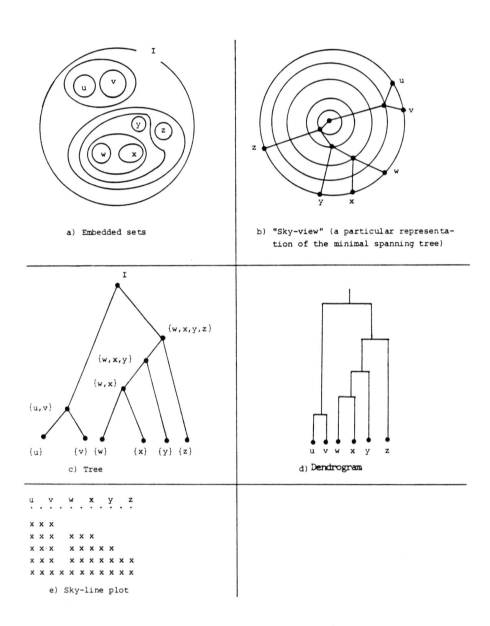

Figure 3.2: Differing representations of a hierarchic clustering on 6 objects.

3.2. MATHEMATICAL DESCRIPTION

gression. As an example, consider the hierarchical classification of galaxies based on certain features. It could be attempted to characterize clusters at higher levels of the tree as being major spiral and elliptical groups, to which other subclasses are related.

The most common interpretation made of hierarchic clustering is to derive a partition: a line is drawn horizontally through the hierarchy, to yield a set of classes. These clusters are precisely the connected components in the case of the single linkage method. A line drawn just above rank 3 (or criterion value 4) on the dendrogram in Figure 3.1 yields classes $\{1,2,4\}$ and $\{3,5\}$. Generally the choice of where "to draw the line" is arrived at on the basis of large changes in the criterion value. However the changes in criterion value increase (usually) towards the final set of agglomerations, which renders the choice of best partition on this basis difficult. Since every line drawn through the dendrogram defines a partition, it may be expedient to choose a partition with convenient features (number of classes, number of objects per class).

A final type of interpretation, less common than the foregoing, is to dispense with the requirement that the classes chosen constitute a partition, and instead detect maximal (i.e. disjoint) clusters of interest at varying levels of the hierarchy. Such an approach is used by Rapoport and Fillenbaum (1972) in a clustering of colours based on semantic attributes.

In summary, a dendrogram provides a résumé of many of the proximity and classificatory relationships in a body of data. It is a convenient representation which answers such questions as: "How many groups are in this data?", "What are the salient interrelationships present?". But it should be stressed that differing answers can feasibly be provided by a dendrogram for most of these questions, just as different human observers would also arrive at different conclusions.

3.2.3 Agglomerative Algorithms

In the last section, a general agglomerative algorithm was discussed. A wide range of these algorithms have been proposed at one time or another. Hierarchic agglomerative algorithms may be conveniently broken down into two groups of methods. The first group is that of linkage methods — the single, complete, weighted and unweighted average linkage methods. These are methods for which a graph representation can be used. Figure 3.3 shows the close relationship between the single linkage hierarchy and the minimal spanning tree; we see that the single linkage dendrogram can be constructed by first sorting the dissimilarities into ascending order; and that in some cases, e.g. in going from d_{bc} to d_{ac}, there is no change in the dendrogram and information which is irrelevant to the minimal spanning tree is ignored. Sneath and Sokal (1973) may be consulted for many other graph representations of the stages in the construction of hierarchic clusterings.

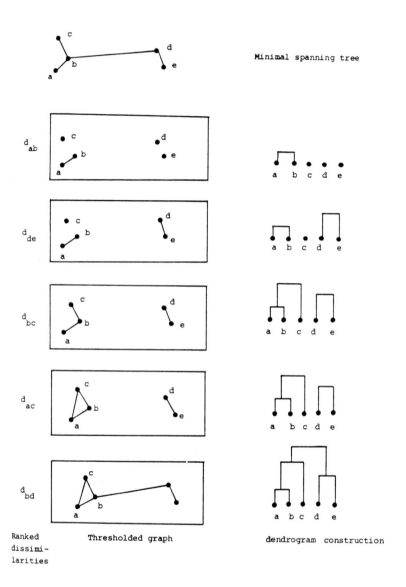

Figure 3.3: Another approach to constructing a single linkage dendrogram.

3.2. MATHEMATICAL DESCRIPTION

The second group of hierarchic clustering methods are methods which allow the cluster centres to be specified (as an average or a weighted average of the member vectors of the cluster). These methods include the centroid, median and minimum variance methods.

The latter may be specified either in terms of dissimilarities, alone, or alternatively in terms of cluster centre coordinates and dissimilarities. A very convenient formulation, in dissimilarity terms, which embraces all the hierarchical methods mentioned so far, is the *Lance–Williams dissimilarity update formula*. If points (objects) i and j are agglomerated into cluster $i \cup j$, then we must simply specify the new dissimilarity between the cluster and all other points (objects or clusters). The formula is:

$$d(i \cup j, k) = \alpha_i d(i,k) + \alpha_j d(j,k) + \beta d(i,j) + \gamma \mid d(i,k) - d(j,k) \mid$$

where α_i, α_j, β, and γ define the agglomerative criterion. Values of these are listed in the second column of Table 3.1. In the case of the single link method, using $\alpha_i = \alpha_j = \frac{1}{2}$, $\beta = 0$, and $\gamma = -\frac{1}{2}$ gives us

$$d(i \cup j, k) = \frac{1}{2} d(i,k) + \frac{1}{2} d(j,k) - \frac{1}{2} \mid d(i,k) - d(j,k) \mid$$

which, it may be verified by taking a few simple examples of three points, i, j, and k, can be rewritten as

$$d(i \cup j, k) = \min \{d(i,k), d(j,k)\}.$$

This was exactly the update formula used in the agglomerative algorithm given in the previous Section. Using other update formulas, as given in Column 2 of Table 3.1, allows the other agglomerative methods to be implemented in a very similar way to the implementation of the single link method.

In the case of the methods which use cluster centres, we have the centre coordinates (in Column 3 of Table 3.1) and dissimilarities as defined between cluster centres (Column 4 of Table 3.1). The Euclidean distance must be used, initially, for equivalence between the two approaches. In the case of the *median method*, for instance, we have the following (cf. Table 3.1).

Let **a** and **b** be two points (i.e. m–dimensional vectors: these are objects or cluster centres) which have been agglomerated, and let **c** be another point. From the Lance–Williams dissimilarity update formula, using squared Euclidean distances, we have:

$$\begin{aligned} d^2(a \cup b, c) &= \frac{d^2(a,c)}{2} + \frac{d^2(b,c)}{2} - \frac{d^2(a,b)}{4} \\ &= \frac{\|\mathbf{a}-\mathbf{c}\|^2}{2} + \frac{\|\mathbf{b}-\mathbf{c}\|^2}{2} - \frac{\|\mathbf{a}-\mathbf{b}\|^2}{4}. \end{aligned} \quad (3.1)$$

The new cluster centre is $(\mathbf{a} + \mathbf{b})/2$, so that its distance to point **c** is

Hierarchical clustering methods (and aliases).	Lance and Williams dissimilarity update formula.	Coordinates of centre of cluster, which agglomerates clusters i and j.	Dissimilarity between cluster centres g_i and g_j.
Single link (nearest neighbour).	$\alpha_i = 0.5$ $\beta = 0$ $\gamma = -0.5$ (More simply: $min\{d_{ik}, d_{jk}\}$)		
Complete link (diameter).	$\alpha_i = 0.5$ $\beta = 0$ $\gamma = 0.5$ (More simply: $max\{d_{ik}, d_{jk}\}$)		
Group average (average link, UPGMA).	$\alpha_i = \frac{|i|}{|i|+|j|}$ $\beta = 0$ $\gamma = 0$		
McQuitty's method (WPGMA).	$\alpha_i = 0.5$ $\beta = 0$ $\gamma = 0$		
Median method (Gower's, WPGMC).	$\alpha_i = 0.5$ $\beta = -0.25$ $\gamma = 0$	$\mathbf{g} = \frac{\mathbf{g}_i + \mathbf{g}_j}{2}$	$\|\mathbf{g}_i - \mathbf{g}_j\|^2$
Centroid (UPGMC).	$\alpha_i = \frac{|i|}{|i|+|j|}$ $\beta = -\frac{|i||j|}{(|i|+|j|)^2}$ $\gamma = 0$	$\mathbf{g} = \frac{|i|\mathbf{g}_i + |j|\mathbf{g}_j}{|i|+|j|}$	$\|\mathbf{g}_i - \mathbf{g}_j\|^2$
Ward's method (minimum variance, error sum of squares.	$\alpha_i = \frac{|i|+|k|}{|i|+|j|+|k|}$ $\beta = -\frac{|k|}{|i|+|j|+|k|}$ $\gamma = 0$	$\mathbf{g} = \frac{|i|\mathbf{g}_i + |j|\mathbf{g}_j}{|i|+|j|}$	$\frac{|i||j|}{|i|+|j|}\|\mathbf{g}_i - \mathbf{g}_j\|^2$

Notes: $|i|$ is the number of objects in cluster i; g_i is a vector in m-space (m is the set of attributes), — either an intial point or a cluster centre; $\|.\|$ is the norm in the Euclidean metric; the names UPGMA, etc. are due to Sneath and Sokal (1973); finally, the Lance and Williams recurrence formula is:

$$d_{i \cup j, k} = \alpha_i d_{ik} + \alpha_j d_{jk} + \beta d_{ij} + \gamma \mid d_{ik} - d_{jk} \mid .$$

Table 3.1: Specifications of seven hierarchical clustering methods.

3.2. MATHEMATICAL DESCRIPTION

$$\|c - \frac{a+b}{2}\|^2. \tag{3.2}$$

That these two expressions are identical is readily verified. The correspondence between these two perspectives on the one agglomerative criterion is similarly proved for the centroid and minimum variance methods.

The single linkage algorithm discussed in the last Section, duly modified for the use of the Lance–Williams dissimilarity update formula, is applicable for all agglomerative strategies. The update formula listed in Table 3.1 is used in Step 2 of the algorithm.

For cluster centre methods, and with suitable alterations for graph methods, the following algorithm is an alternative to the general dissimilarity based algorithm (the latter may be described as a "stored dissimilarities approach").

Stored data approach

Step 1 Examine all interpoint dissimilarities, and form cluster from two closest points.

Step 2 Replace two points clustered by representative point (centre of gravity) or by cluster fragment.

Step 3 Return to Step 1, treating clusters as well as remaining objects, until all objects are in one cluster.

In Steps 1 and 2, "point" refers either to objects or clusters, both of which are defined as vectors in the case of cluster centre methods. This algorithm is justified by storage considerations, since we have $O(n)$ storage required for n initial objects and $O(n)$ storage for the $n-1$ (at most) clusters. In the case of linkage methods, the term "fragment" in Step 2 refers (in the terminology of graph theory) to a connected component in the case of the single link method and to a clique or complete subgraph in the case of the complete link method. The overall complexity of the above algorithm is $O(n^3)$: the repeated calculation of dissimilarities in Step 1, coupled with $O(n)$ iterations through Steps 1, 2 and 3. Note however that this does not take into consideration the extra processing required in a linkage method, where "closest" in Step 1 is defined with respect to graph fragments.

Recently some other very efficient improvements on this algorithm have been proposed (for a survey, see Murtagh, 1985). In particular there is the *Nearest Neighbour (NN) chain* algorithm. Here is a short description of this approach.

A NN–chain consists of an arbitrary point (a in Figure 3.4); followed by its NN (b in Figure 3.4); followed by the NN from among the remaining points (c, d, and e in Figure 3.4) of this second point; and so on until we necessarily have

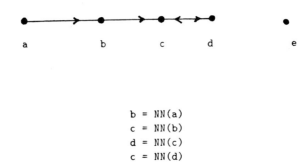

Figure 3.4: Five points, showing NNs and RNNs.

some pair of points which can be termed reciprocal or mutual NNs. (Such a pair of RNNs may be the first two points in the chain; and we have assumed that no two dissimilarities are equal.)

In constructing a NN–chain, irrespective of the starting point, we may agglomerate a pair of RNNs as soon as they are found. What guarantees that we can arrive at the same hierarchy as if we used the "stored dissimilarities" or "stored data" algorithms described earlier in this section? Essentially this is the same condition as that under which no inversions or reversals are produced by the clustering method. Figure 3.5 gives an example of this, where d is agglomerated at a lower criterion value (i.e. dissimilarity) than was the case at the previous agglomeration.

This is formulated as:

Inversion impossible if : $d(i,j) < d(i,k)$ or $d(j,k) \Rightarrow d(i,j) < d(i \cup j, k)$

Using the Lance–Williams dissimilarity update formula, it can be shown that the minimum variance method does not give rise to inversions; neither do the linkage methods; but the median and centroid methods cannot be guaranteed not to have inversions.

To return to Figure 3.4, if we are dealing with a clustering criterion which precludes inversions, then c and d can justifiably be agglomerated, since no other point (for example, b or e) could have been agglomerated to either of these.

The processing required, following an agglomeration, is to update the NNs of points such as b in Figure 3.4 (and on account of such points, this algorithm was initially dubbed *algorithme des célibataires* when first proposed!). The following is a summary of the algorithm:

3.2. MATHEMATICAL DESCRIPTION   67

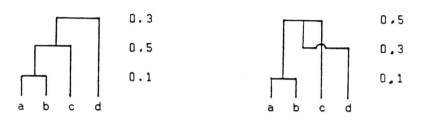

Figure 3.5: Alternative representations of a hierarchy with an inversion.

NN–chain algorithm

Step 1 Select a point arbitrarily.

Step 2 Grow the NN–chain from this point until a pair of RNNs are obtained.

Step 3 Agglomerate these points (replacing with a cluster point, or updating the dissimilarity matrix).

Step 4 From the point which preceded the RNNs (or from any other arbitrary point if the first two points chosen in Steps 1 and 2 constituted a pair of RNNs), return to Step 2 until only one point remains.

3.2.4 Minimum Variance Method in Perspective

The next Section will review mathematical properties of the minimum variance method; in this Section we will informally motivate our preference for this method. First, let us briefly review the overall perspective.

Agglomerative clustering methods have been motivated by graph theory (leading to linkage–based methods) or by geometry (leading to cluster centre methods). This is true for the more commonly used methods studied here. In cluster centre methods, the cluster centre may be used for subsequent agglomerations. Alternatively, inter–cluster dissimilarities may be used throughout, and therefore these methods may be implemented using the Lance–Williams dissimilarity update formula (see Table 3.1) in an identical manner to the linkage–based methods.

In order to specify an agglomerative criterion simultaneously in terms of cluster mean vectors, and in terms of dissimilarity, it was also necessary to adopt a particular dissimilarity (i.e. the Euclidean distance). Restricting the choice of dissimilarity to this distance is not usually inconvenient in practice, and a Euclidean space offers a well–known and powerful standpoint for analysis.

The variance or spread of a set of points (i.e. the sum of squared distances from the centre) has been the point of departure for specifying many clustering algorithms. Many of these algorithms, — iterative, optimization algorithms as well as the hierarchical, agglomerative algorithms — are briefly described and appraised in Wishart (1969). The use of variance in a clustering criterion links the resulting clustering to other data–analytic techniques which involve a decomposition of variance. Principal Components Analysis, for example, which has been studied in Chapter 2 seeks the principal directions of elongation of the multidimensional points, i.e. the axes on which the projections of the points have maximal variance. Using a clustering of the points with minimal variance within clusters as the cluster criterion is, perhaps, the most suitable criterion for two different but complimentary analyses of the same set of points. The reality of clusters of projected points resulting from the Principal Components Analysis may be assessed using the Cluster Analysis results; and the interpretation of the axes of the former technique may be used to facilitate interpretation of the clustering results.

The search for clusters of maximum homogeneity leads to the minimum variance criterion. Since no coordinate axis is privileged by the Euclidean distance, the resulting clusters will be approximately hyperspherical. Such ball–shaped clusters will therefore be very unsuitable for examining straggly patterns of points. However, in the absence of information about such patterns in the data, homogeneous clusters will provide the most useful condensation of the data.

The following properties make the minimum variance agglomerative strategy particularly suitable for synoptic clustering:

1. As discussed in the Section to follow, the two properties of cluster homogeneity and cluster separability are incorporated in the cluster criterion. For summarizing data, it is unlikely that more suitable criteria could be devised.

2. As in the case of other geometric strategies, the minimum variance method defines a cluster centre of gravity. This mean set of cluster members' coordinate values is the most useful summary of the cluster. It may also be used for the fast selection and retrieval of data, by matching on these cluster representative vectors rather than on each individual object vector.

3. A top–down hierarchy traversal algorithm may also be implemented for information retrieval. Using a query vector, the left or right subtree is selected

3.2. MATHEMATICAL DESCRIPTION

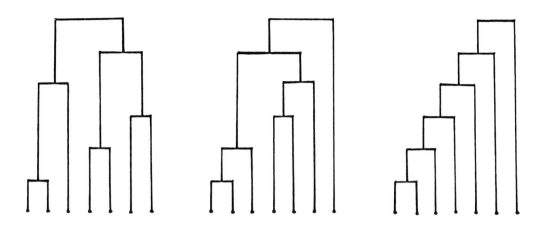

Figure 3.6: Three binary hierarchies: symmetric, asymmetric and intermediate.

at each node for continuation of the traversal (it is best to ensure that each node has precisely two successor nodes in the construction of the hierarchy). Such an algorithm will work best if all top–down traversals through the hierarchy are of approximately equal length. This will be the case if and only if the hierarchy is as "symmetric" or "balanced" as possible (see Figure 3.6). Such a balanced hierarchy is usually of greatest interest for interpretative purposes also: a partition, derived from a hierarchy, and consisting of a large number of small classes, and one or a few large classes, is less likely to be of practical use.

For such reasons, a "symmetric" hierarchy is desirable. It has been shown, using a number of different measures of hierarchic symmetry, that the minimum variance (closely followed by the complete link) methods generally give the most symmetric hierarchies (see Murtagh, 1984).

4. Unlike other geometric agglomerative methods — in particular the centroid and the median methods (see definitions, Table 3.1, above) — the sequence of agglomerations in the minimum variance method is guaranteed not to allow inversions in the cluster criterion value. Inversions or reversals (Figure 3.5) are inconvenient, and can make interpretation of the hierarchy difficult.

5. Finally, computational performance has until recently favoured linkage based agglomerative criteria, and in particular the single linkage method. The computational advances described above for the minimum variance method (principally the NN–chain algorithm) make it increasingly attractive for

practical applications involving large amounts of data.

3.2.5 Minimum Variance Method: Mathematical Properties

The minimum variance method produces clusters which satisfy compactness and isolation criteria. These criteria are incorporated into the dissimilarity, noted in Table 3.1, as will now be shown.

In Ward's method, we seek to agglomerate two clusters, c_1 and c_2, into cluster c such that the within–class variance of the partition thereby obtained is minimum. Alternatively, the between–class variance of the partition obtained is to be maximized. Let P and Q be the partitions prior to, and subsequent to, the agglomeration; let $p_1, p_2, \ldots$ be classes of the partitions:

$$P = \{p_1, p_2, \ldots, p_k, c_1, c_2\}$$
$$Q = \{p_1, p_2, \ldots, p_k, c\}.$$

Finally, let i denote any individual or object, and I the set of such objects. In the following, classes (i.e. p or c) and individuals (i.e. i) will be considered as vectors or as sets: the context, and the block typing of vectors, will be sufficient to make clear which is the case.

Total variance of the cloud of objects in m–dimensional space is decomposed into the sum of within–class variance and between–class variance. This is Huyghen's theorem in classical mechanics. Let V denote *variance*. The total variance of the cloud of objects is

$$V(I) = \frac{1}{n} \sum_{i \in I} (\mathbf{i} - \mathbf{g})^2$$

where $\mathbf{g}$ is the grand mean of the n objects: $\mathbf{g} = \frac{1}{n} \sum_{i \in I} \mathbf{i}$. The between–class variance is

$$V(P) = \sum_{p \in P} \frac{|p|}{n} (\mathbf{p} - \mathbf{g})^2$$

where $|p|$ is the cardinality of (i.e. number of members in) class p. (Note that $\mathbf{p}$ — in block type–face — is used to denote the centre of gravity — a vector — and p the set whose centre of gravity this is). Finally, the within–class variance is

$$\frac{1}{n} \sum_{p \in P} \sum_{i \in p} (\mathbf{i} - \mathbf{p})^2.$$

For two partitions, before and after an agglomeration, we have respectively:

3.2. MATHEMATICAL DESCRIPTION

$$V(I) = V(P) + \sum_{p \in P} V(p)$$

$$V(I) = V(Q) + \sum_{p \in Q} V(p).$$

Hence,

$$V(P) + V(p_1) + \ldots + V(p_k) + V(c_1) + V(c_2)$$
$$= V(Q) + V(p_1) + \ldots + V(p_k) + V(c).$$

Therefore:

$$V(Q) = V(P) + V(c_1) + V(c_2) - V(c).$$

In agglomerating two classes of P, the variance of the resulting partition (i.e. $V(Q)$) will necessarily decrease: therefore in seeking to minimize this decrease, we simultaneously achieve a partition with maximum between–class variance. The criterion to be optimized can then be shown to be:

$$\begin{aligned} V(P) - V(Q) &= V(c) - V(c_1) - V(c_2) \\ &= \tfrac{|c_1|\,|c_2|}{|c_1|+|c_2|} \|\mathbf{c_1} - \mathbf{c_2}\|^2 \,, \end{aligned}$$

which is the dissimilarity given in Table 3.1. This is a dissimilarity which may be determined for any pair of classes of partition P; and the agglomerands are those classes, c_1 and c_2, for which it is minimum.

It may be noted that if c_1 and c_2 are singleton classes, then $V(\{c_1, c_2\}) = \tfrac{1}{2}\|\mathbf{c_1} - \mathbf{c_2}\|^2$ (i.e. the variance of a pair of objects is equal to half their Euclidean distance).

3.2.6 Minimal Spanning Tree

Aspects of the minimal spanning tree (MST) are covered in most texts on graph theory, and on many other areas besides. In graph theory — a subdiscipline of discrete mathematics — *vertices* (i.e. points) and *edges* (i.e. lines joining the points) are commonly dealt with. A graph is defined as a set of such vertices and edges. A MST contains (i.e. spans) all vertices, and has minimal totalled edge length (or *weight*). We will restrict ourselves to a brief description of this important method. The following algorithm formalises Figure 3.3 in constructing a MST by a "greedy" or nearest neighbour approach.

Minimal Spanning Tree algorithm

Step 1 Select an arbitrary point and connect it to the least dissimilar neighbour. These two points constitute a subgraph of the MST.

Step 2 Connect the current subgraph to the least dissimilar neighbour of any of the members of the subgraph.

Step 3 Loop on Step 2, until all points are in the one subgraph: this, then, is the MST.

Step 2 agglomerates subsets of objects using the criterion of connectivity. For proof that this algorithm does indeed produce a MST, see for example Tucker (1980). The close relationship between the MST and the single linkage hierarchical clustering method is illustrated in Figure 3.3.

A *component* in a graph is a subgraph consisting of a set of vertices with at least one edge connecting each vertex to some other vertex in the component. Hence a component is simply one possible definition of a cluster, — in fact, a component is closely related to clusters which can be derived from a single linkage hierarchy (cf. Figure 3.3).

Breaking up the MST, and thereby automatically obtaining components, is a problem addressed by Zahn (1971). He defined an edge to be *inconsistent* if it is of length much greater than the lengths of other nearby edges (see Figure 3.7).

Inconsistent edges may be picked out by looking at a histogram of edge lengths in the MST, and marking as deletable a set percentage of greatest–length edges. Alternatively, inconsistent edges may be obtained by defining a threshold of inconsistency: if it is supposed, for instance, that edge lengths are normally distributed, a threshold of two standard deviations above the mean length of all edges which are within two edges (say, for some robustness) of the given vertex may be a useful indicator of inconsistency. Rarely is it worthwhile to attempt to test a supposition such as normality of edge lengths, and instead a rule such as this would be judged only on the results given in practice. Zahn applied these approaches to point pattern recognition, — obtaining what he termed "Gestalt patterns" among sets of planar points (see Fig. 3.8); picking out bubble chamber particle tracks, indicated by curved sequences of points; and detecting density gradients, where differing clusters of points have different densities associated with them and hence are distinguishable to the human eye. The MST provides a useful starting point for undertaking such pattern recognition problems.

3.2. MATHEMATICAL DESCRIPTION

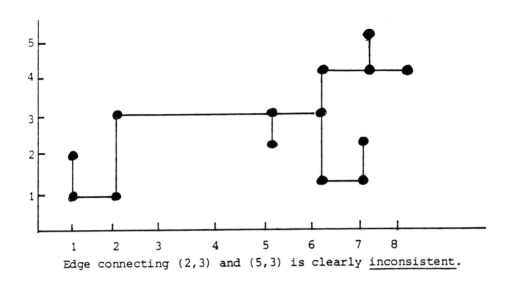

Figure 3.7: Minimal spanning tree of a point pattern (non-unique).

The MST is also often suitable for outlier detection. Since outlying data items will be of greater than average distance from their neighbours in the MST, they may be detected by drawing a histogram of edge lengths. Unusually large lengths will indicate the abnormal points (or data items) sought. Rohlf (1975) gave a statistical gap test, under the assumption that the edge lengths in the MST were normally distributed.

3.2.7 Partitioning Methods

We will conclude this Chapter with a short look at other non–hierarchical clustering methods.

A large number of assignment algorithms have been proposed. The single–pass approach usually achieves computational efficiency at the expense of precision, and there are many iterative approaches for improving on crudely–derived partitions.

As an example of a single–pass algorithm, the following one is given in Salton and McGill (1983). The general principle followed is: make one pass through the data, assigning each object to the first cluster which is close enough, and making a new cluster for objects that are not close enough to any existing cluster.

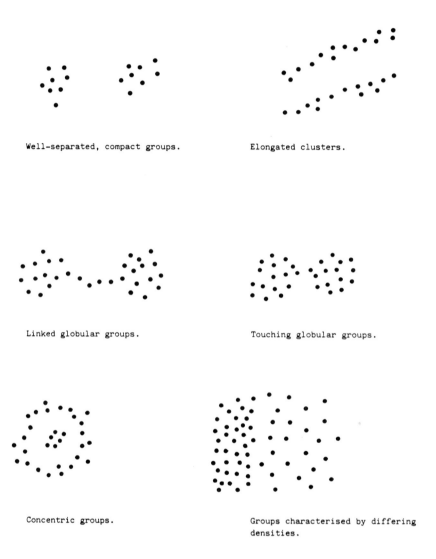

Figure 3.8: Differing point patterns.

3.2. MATHEMATICAL DESCRIPTION

Single–pass overlapping cluster algorithm

Input n objects, threshold t, dissimilarity on objects.

Step 1 Read object 1, and insert object 1 in membership list of cluster 1. Let representative of cluster 1 be given by object 1. Set i to 2.

Step 2 Read i^{th} object. If $diss(\ i^{th}$ object, cluster $j\ ) \leq t$, for any cluster j, then include the ith object in the membership list of cluster j, and update the cluster representative vector to take account of this new member. If $diss(\ i^{th}$ object, cluster $j) > t$, for all clusters j, then create a new cluster, placing the ith object in its membership list, and letting the representative of this cluster be defined by the ith object.

Step 3 Set i to $i+1$. If $i \leq n$, go to Step 2.

The cluster representative vector used is usually the mean vector of the cluster's members; in the case of binary data, this representative might then be thresholded to have 0 and 1 coordinate values only. In Step 2, it is clear that overlapping clusters are possible in the above algorithm. In the worst case, if threshold t is chosen too low, all n objects will constitute clusters and the number of comparisons to be carried out will be $O(n^2)$. The dependence of the algorithm on the given sequence of objects is an additional disadvantage of this algorithm. However, its advantages are that it is conceptually very simple, and for a suitable choice of threshold will probably not require large processing time. In practice, it can be run for a number of different values of t.

As a non-hierarchic strategy, it is hardly surprising that the variance criterion has always been popular (for some of the same reasons as were seen in Section 3.2.4 for the hierarchical approach based on this criterion). We may, for instance, minimize the within–class variance

$$V_{opt} = min_P \sum_{p \in P} \sum_{i \in p} \|\mathbf{i} - \mathbf{p}\|^2$$

where the partition P consists of classes p of centre $\mathbf{p}$, and we desire the minimum of this criterion over all possible partitions, P. To avoid a nontrivial outcome (e.g. each class being singleton, giving zero totalled within class variance), the number of classes (k) must be set.

A hierarchical clustering, using the minimum variance criterion, then provides a solution, — not necessarily optimal — at level $n - k$ when n objects are being processed. An alternative approach uses iterative refinement, as follows.

Iterative optimization algorithm for the variance criterion

Step 1 Arbitrarily define a set of k cluster centres.

Step 2 Assign each object to the cluster to which it is closest (using the Euclidean distance, $d^2(i,p) = \|\mathbf{i} - \mathbf{p}\|^2$).

Step 3 Redefine cluster centres on the basis of the current cluster memberships.

Step 4 If the totalled within class variances is better than at the previous iteration, then return to Step 2.

We have omitted in Step 4 a test for convergence (the number of iterations should not exceed, e.g., 25). Cycling is also possible between solution states. This algorithm could be employed on the results (at level $n-k$) of a hierarchic clustering in order to improve the partition found. It is however a suboptimal algorithm, — the minimal distance strategy used by this algorithm is clearly a *sufficient* but not a *necessary* condition for an optimal partition.

The initial cluster centres may be chosen arbitrarily (for instance, by averaging a small number of object–vectors); or they may be chosen from prior knowledge of the data. In the latter case we may for example "manually" choose a small set of stars and galaxies, determine their (parameter–space) centres of gravity, and use these as the basis for the classification of objects derived from a digitized image.

Yet another approach to optimizing the same minimum variance criterion is the exchange method.

Exchange method for the minimum variance criterion

Step 1 Arbitrarily choose an initial partition.

Step 2 For each $i \in p$, see if the criterion is bettered by relocating i in another class q. If this is the case, we choose class q such that the criterion V is least; if it is not the case, we proceed to the next i.

Step 3 If the maximum possible number of iterations has not been reached, and if at least one relocation took place in Step 2, return again to Step 2.

Two remarks may be made: we will normally require that an object not be removed from a singleton class; and the change in variance brought about by relocating object i from class p to class q can be shown to be

$$\frac{|p|}{|p|-1}\|\mathbf{i}-\mathbf{p}\|^2 - \frac{|q|}{|q|-1}\|\mathbf{i}-\mathbf{q}\|^2$$

Hence, if this expression is positive, i ought to be relocated from class p (to class q if another, better, class is not found).

Späth (1985) offers a lucid and thorough treatment of algorithms of the sort described.

In terminating, it is necessary to make some suggestion as to when these partitioning algorithms should be used in preference to hierarchical algorithms. We have seen that the number of classes must be specified, as also the requirement that each class be non-empty. A difficulty with iterative algorithms, in general, is the requirement for parameters to be set in advance (Anderberg, 1973, describes a version of the ISODATA iterative clustering method which requires 7 pre–set parameters). As a broad generalization, it may thus be asserted that iterative algorithms ought to be considered when the problem is clearly defined in terms of numbers and other characteristics of clusters; but hierarchical routines often offer a more general–purpose and user–friendly option.

3.3 Examples and Bibliography

3.3.1 Examples from Astronomy

Principal Components Analysis has often been used for determining a classification, and these references are not included in this Section (see Section 2.3.3).

The problems covered in the following include: star-galaxy separation, using digitized image data; spectral classification, — the prediction of spectral type from photometry; taxonomy construction (for asteroids, stars, and stellar light curves); galaxies; gamma and X–ray astronomy; a clustering approach not widely used elsewhere is employed for studies relating to the Moon, to asteroids and to cosmic sources; and work relating to interferogram analysis is represented.

1. J.D. Barrow, S.P. Bhavsar and D.H. Sonoda, "Minimal spanning trees, filaments and galaxy clustering", *Monthly Notices of the Royal Astronomical Society*, **216**, 17–35, 1985.

   (This article follows the seminal approach of Zahn — see reference among the general clustering works — in using the MST for finding visually evident groupings.)

2. R. Bianchi, A. Coradini and M. Fulchignoni, "The statistical approach to the study of planetary surfaces", *The Moon and the Planets*, **22**, 293–304, 1980.

   (This article contains a general discussion which compares the so–called G–mode clustering method to other multivariate statistical methods. References 7 and 8 below also use this method.)

3. R. Bianchi, J.C. Butler, A. Coradini and A.I. Gavrishin, "A classification of lunar rock and glass samples using the G–mode central method", *The Moon*

and the Planets, **22**, 305–322, 1980.

4. A. Bijaoui, "Méthodes mathématiques pour la classification stellaire", in *Classification Stellaire, Compte Rendu de l'Ecole de Goutelas*, ed. D. Ballereau, Observatoire de Meudon, Meudon, 1979, pp. 1–54.

   (This presents a survey of clustering methods.)

5. R. Buccheri, P. Coffaro, G. Colomba, V. Di Gesù and S. Salemi, "Search of significant features in a direct non–parametric pattern recognition method. Application to the classification of multiwire spark chamber pictures", in (eds.) C. de Jager and H. Nieuwenhuijzen, *Image Processing Techniques in Astronomy*, D. Reidel, Dordrecht, pp. 397–402, 1975.

   (A technique is developed for classifying γ–ray data.)

6. S.A. Butchins, "Automatic image classification", *Astronomy and Astrophysics*, **109**, 360–365, 1982.

   (A method for determining Gaussian clusters, due to Wolf, is used for star/galaxy separation in photometry.)

7. A. Coradini, M. Fulchignoni and A.I. Gavrishin, "Classification of lunar rocks and glasses by a new statistical technique", *The Moon*, **16**, 175–190, 1976.

   (The above, along with the references of Bianchi and others, make use of a clustering technique called the G–mode method. The above contains a short mathematical description of the technique proposed.)

8. A. Carusi and E. Massaro, "Statistics and mapping of asteroid concentrations in the proper elements' space", *Astronomy and Astrophysics Supplement Series*, **34**, 81–90, 1978.

   (This article also uses the so–called G–mode method, employed by Bianchi, Coradini, and others.)

9. C.R. Cowley and R. Henry, "Numerical taxonomy of Ap and Am stars", *The Astrophysical Journal*, **233**, 633–643, 1979.

   (40 stars are used, characterised on the strength with which particular atomic spectra — the second spectra of yttrium, the lanthanides, and the iron group — are represented in the spectrum. Stars with very similar spectra end up correctly grouped; and anomolous objects are detected. Clustering using lanthanides, compared to clustering using iron group data, gives different results for A_p stars. This is not the case for A_m stars, which thus appear to be less heterogeneous. The need for physical explanations are thus suggested.)

3.3. EXAMPLES AND BIBLIOGRAPHY

10. C.R. Cowley, "Cluster analysis of rare earths in stellar spectra", in *Statistical Methods in Astronomy*, European Space Agency Special Publication 201, 1983, pp. 153–156.

    (About twice the number of stars, as used in the previous reference, are used here. A greater role is seen for chemical explanations of stellar abundances and/or spectroscopic patterns over nuclear hypotheses.)

11. J.K. Davies, N. Eaton, S.F. Green, R.S. McCheyne and A.J. Meadows, "The classification of asteroids", *Vistas in Astronomy*, **26**, 243–251, 1982.

    (Physical properties of 82 asteroids are used. The dendrogram obtained is compared with other classification schemes based on spectral characteristics or colour–colour diagrams. The clustering approach used is justified also in being able to pinpoint objects of particular interest for further observation; and in allowing new forms of data — e.g. broadband infrared photometry — to be quickly incorporated into the overall approach of classification-construction.)

12. G.A. De Biase, V. di Gesù and B. Sacco, "Detection of diffuse clusters in noise background", *Pattern Recognition Letters* **4**, 39–44, 1986.

13. P.A. Devijver, "Cluster analysis by mixture identification", in V. Di Gesù, L. Scarsi, P. Crane, J.H. Friedman and S. Levialdi (eds.), *Data Analysis in Astronomy*, Plenum Press, New York, 1984, pp. 29–44.

    (A useful review article, with many references. A perspective similar to perspectives adopted by many discriminant analysis methods is used.)

14. V. Di Gesù and B. Sacco, "Some statistical properties of the minimum spanning forest", *Pattern Recognition*, **16**, 525–531, 1983.

    (In this and the following works, the minimal spanning tree or fuzzy set theory — which, is clear from the article titles — are applied to point pattern distinguishing problems involving gamma and X-ray data. For a rejoinder to the foregoing reference, see R.C. Dubes and R.L. Hoffman, "Remarks on some statistical properties of the minimum spanning forest", *Pattern Recognition*, **19**, 49–53, 1986. A reply to this article is forthcoming, from the authors of the original paper.)

15. V. Di Gesù, B. Sacco and G. Tobia, "A clustering method applied to the analysis of sky maps in gamma-ray astronomy", *Memorie della Società Astronomica Italiana*, 517–528, 1980.

16. V. Di Gesù and M.C. Maccarone, "A method to classify celestial shapes based on the possibility theory", in G. Sedmak (ed.), ASTRONET 1983

(Convegno Nazionale Astronet, Brescia, Published under the auspices of the Italian Astronomical Society), 355–363, 1983.

17. V. Di Gesù and M.C. Maccarone, "Method to classify spread shapes based on possibility theory", Proceedings of the 7th International Conference on Pattern Recognition, Vol. 2, IEEE Computer Society, 1984, pp. 869–871.

18. V. Di Gesù and M.C. Maccarone, "Features selection and possibility theory", Pattern Recognition, 19, 63–72, 1986.

19. J.V. Feitzinger and E. Braunsfurth, "The spatial distribution of young objects in the Large Magellanic Cloud — a problem of pattern recognition", in eds. S. van den Bergh and K.S. de Boer, Structure and Evolution of the Magellanic Clouds, IAU, 93–94, 1984.

(In an extended abstract, the use of linkages between objects is described.)

20. I.E. Frank, B.A. Bates and D.E. Brownlee, "Multivariate statistics to analyze extraterrestrial particles from the ocean floor", in V. Di Gesù, L. Scarsi, P. Crane, J.H. Friedman and S. Levialdi (eds.), Data Analysis in Astronomy, Plenum Press, New York, 1984.

21. A. Fresneau, "Clustering properties of stars outside the galactic disc", in Statistical Methods in Astronomy, European Space Agency Special Publication 201, 1983, pp. 17–20.

(Techniques from the spatial processes area of statistics are used to assess clustering tendencies of stars.)

22. A. Heck, A. Albert, D. Defays and G. Mersch, "Detection of errors in spectral classification by cluster analysis", Astronomy and Astrophysics, 61, 563–566, 1977.

23. A. Heck, D. Egret, Ph. Nobelis and J.C. Turlot, "Statistical confirmation of the UV spectral classification system based on IUE low–dispersion stellar spectra", Astrophysics and Space Science, 120, 223–237, 1986.

(Among other results, it is found that UV standard stars are located in the neighbourhood of the centres of gravity of groups found, thereby helping to verify the algorithm implemented. A number of other papers, by the same authors, analysing IUE spectra are referenced in this paper. Apart from the use of a large range of clustering methods, these papers also introduce a novel weighting procedure — termed the "variable Procrustean bed" (see Chapter 6) — which adjusts for the symmetry/asymmetry of the spectrum. Therefore, a useful study of certain approaches to the coding of data is to be found in these papers.)

3.3. EXAMPLES AND BIBLIOGRAPHY

24. J.P. Huchra and M.J. Geller, "Groups of galaxies. I. Nearby groups", *The Astrophysical Journal*, **257**, 423–437, 1982.

    (The single linkage hierarchical method, or the minimal spanning tree, have been rediscovered many times — see, for instance, Graham and Hell, 1985, referenced in the general clustering section. In this study, a close variant is used for detecting groups of galaxies using three variables, — two positional variables and redshift.)

25. J.F. Jarvis and J.A. Tyson, "FOCAS: faint object classification and analysis system", *The Astronomical Journal*, **86**, 476–495, 1981.

    (An iterative minimal distance partitioning method is employed in the FOCAS system to arrive at star/galaxy/other classes.)

26. G. Jasniewicz, "The Böhm–Vitense gap in the Geneva photometric system", *Astronomy and Astrophysics*, *141*, 116–126, 1984.

    (The minimal spanning tree is used on colour–colour diagrams.)

27. A. Kruszewski, "Object searching and analyzing commands", in *MIDAS — Munich Image Data Analysis System*, European Southern Observatory Operating Manual No. 1, Chapter 11, 1985.

    (The *Inventory* routine in MIDAS has a non–hierarchical iterative optimization algorithm. It can immediately work on up to 20 parameters, determined for each object in a scanned image.)

28. M.J. Kurtz, "Classification methods: an introductory survey", in *Statistical Methods in Astronomy*, European Space Agency Special Publication 201, 1983, pp. 47–58.

    (Kurtz lists a large number of parameters — and functions of these parameters — which have been used to differentiate stars from galaxies.)

29. J. Materne, "The structure of nearby clusters of galaxies. Hierarchical clustering and an application to the Leo region", *Astronomy and Astrophysics*, **63**, 401–409, 1978.

    (Ward's minimum variance hierarchic method is used, following discussion of the properties of other hierarchic methods.)

30. M.O. Mennessier, "A cluster analysis of visual and near–infrared light curves of long period variable stars", in *Statistical Methods in Astronomy*, European Space Agency Special Publication 201, 1983, pp. 81–84.

    (Light curves — the variation of luminosity with time in a wavelength range — are analysed. Standardization is applied, and then three hierarchical methods. "Stable clusters" are sought from among all of these. The study is continued in the following.)

31. M.O. Mennessier, "A classification of miras from their visual and near–infrared light curves: an attempt to correlate them with their evolution", *Astronomy and Astrophysics*, **144**, 463–470, 1985.

32. MIDAS (Munich Image Data Analysis System), European Southern Observatory, Garching–bei–München (Version 4.1, January 1986). Chapter 13: Multivariate Statistical Methods.

    (This premier astronomical data reduction package contains a large number of multivariate algorithms.)

33. M. Moles, A. del Olmo and J. Perea, "Taxonomical analysis of superclusters. I. The Hercules and Perseus superclusters", *Monthly Notices of the Royal Astronomical Society*, **213**, 365–380, 1985.

    (A non–hierarchical descending method, used previously by Paturel, is employed.)

34. F. Murtagh, "Clustering techniques and their applications", *Data Analysis and Astronomy* (Proceedings of International Workshop on Data Analysis and Astronomy, Erice, Italy, April 1986) Plenum Press, New York, 1986 (in press). ESO Scientific Preprint No. 448.

35. F. Murtagh and A. Lauberts, "A curve matching problem in astronomy", *Pattern Recognition Letters*, 1986 (in press). ESO Scientific Preprint No. 471.

    (A dissimilarity is defined between galaxy luminosity profiles, in order to arrive at a spiral–elliptical grouping.)

36. G. Paturel, "Etude de la région de l'amas Virgo par taxonomie", *Astronomy and Astrophysics*, **71**, 106–114, 1979.

    (A descending non–hierarchical method is used.)

37. J. Perea, M. Moles and A. del Olmo, "Taxonomical analysis of the Cancer cluster of galaxies", *Monthly Notices of the Royal Astronomical Society*, **222**, 49–53, 1986.

    (A non–hierarchical descending method is used.)

38. D.J. Tholen, "Asteroid taxonomy from cluster analysis of photometry", PhD Thesis, University of Arizona, 1984.

    (Between 400 and 600 asteroids using good–quality multi–colour photometric data are analysed.)

39. F. Giovannelli, A. Coradini, J.P. Lasota and M.L. Polimene, "Classification of cosmic sources: a statistical approach", *Astronomy and Astrophysics*, **95**, 138–142, 1981.

40. B. Pirenne, D. Ponz and H. Dekker, "Automatic analysis of interferograms", *The Messenger*, No. 42, 2–3, 1985.

    (The minimal spanning tree is used to distinguish fringes; there is little description of the MST approach in the above article, but further articles are in preparation and the software — and accompanying documentation — are available in the European Southern Observatory's MIDAS image processing system.)

41. A. Zandonella, " Object classification: some methods of interest in astronomical image analysis", in *Image Processing in Astronomy*, eds. G. Sedmak, N. Capacciolí and R.J. Allen, Osservatorio Astronomico di Trieste, Trieste, 304–318, 1979.

    (This presents a survey of clustering methods.)

3.3.2 General References

1. M.R. Anderberg, *Cluster Analysis for Applications*, Academic Press, New York, 1973.

    (A little dated, but still very much referenced; good especially for similarities and dissimilarities.)

2. J.P. Benzécri et coll., *L'Analyse des Données. I. La Taxinomie*, Dunod, Paris, 1979 (3rd ed.).

    (Very influential in the French speaking world; extensive treatment, and impressive formalism.)

3. R.K. Blashfield and M.S. Aldenderfer, "The literature on cluster analysis", *Multivariate Behavioral Research*, **13**, 271–295, 1978.

4. H.H. Bock, *Automatische Klassifikation*, Vandenhoek und Rupprecht, Göttingen, 1974.

    (Encyclopaedic.)

5. CLUSTAN, Clustan Ltd., 16 Kingsburgh Road, Edinburgh EH12 6DZ, Scotland.

    (The only exclusively clustering package available.)

6. B. Everitt, *Cluster Analysis*, Heinemann Educational Books, London, 1980 (2nd ed.).

(A readable, introductory text.)

7. A.D. Gordon, *Classification*, Chapman and Hall, London, 1981.

   (Another recommendable introductory text.)

8. R.L. Graham and P. Hell, "On the history of the minimum spanning tree problem", *Annals of the History of Computing*, **7**, 43–57, 1985.

   (An interesting historical study.)

9. J.A. Hartigan, *Clustering Algorithms*, Wiley, New York, 1975.

   (Often referenced, this book could still be said to be innovative in its treatment of clustering problems; it contains a wealth of sample data sets.)

10. M. Jambu and M.O. Lebeaux, *Cluster Analysis and Data Analysis*, North-Holland, Amsterdam, 1983.

    (Some of the algorithms discussed have been overtaken by, for instance, the "nearest neighbour chain" or "reciprocal nearest neighbour" algorithms.)

11. L. Lebart, A. Morineau and K.M. Warwick, *Multivariate Descriptive Statistical Analysis*, Wiley, New York, 1984.

    (A useful book, centred on Multiple Correspondence Analysis.)

12. R.C.T. Lee, "Clustering analysis and its applications", in J.T. Tou (ed.) *Advances in Information Systems Science, Vol. 8*, Plenum Press, New York, 1981, pp. 169–292.

    (Practically book-length, it is especially useful for the links between clustering and problems in computing and in Operations Research.)

13. F. Murtagh, "Structure of hierarchic clusterings: implications for information retrieval and for multivariate data analysis", *Information Processing and Management*, **20**, 611–617, 1984.

14. F. Murtagh, *Multidimensional Clustering Algorithms*, COMPSTAT Lectures Volume 4, Physica-Verlag, Wien, 1985.

    (Algorithmic details of a range of clustering methods.)

15. A. Rapoport and S. Fillenbaum, "An experimental study of semantic structures", in eds. A.K. Romney, R.N. Shepard and S.B. Nerlove, *Multidimensional Scaling; Theory and Applications in the Behavioral Sciences. Vol. 2, Applications*, Seminar Press, New York, 93–131, 1972.

16. F.J. Rohlf, "Generalization of the gap test for the detection of multivariate outliers", *Biometrics*, **31**, 93–101, 1975.

    (One application of the Minimal Spanning Tree.)

17. G. Salton and M.J. McGill, *Introduction to Modern Information Retrieval*, McGraw–Hill, New York, 1983.

   (A central reference in the information retrieval area.)

18. P.H.A. Sneath and R.R. Sokal, *Numerical Taxonomy*, Freeman, San Francisco, 1973.

   (Very important for biological applications, it also has some impressive collections of graph representations of clustering results.)

19. H. Späth, *Cluster Dissection and Analysis: Theory, Fortran Programs, Examples*, Ellis Horwood, Chichester, 1985.

   (Recommendable reference for non–hierarchic, partitioning methods.)

20. A. Tucker, *Applied Combinatorics*, Wiley, New York, 1980.

   (For graph theory and combinatorics.)

21. D. Wishart, "Mode analysis: a generalization of nearest neighbour which reduces chaining effects", in ed. A.J. Cole, *Numerical Taxonomy*, Academic Press, New York, 282–311, 1969.

   (Discusses various variance–based clustering criteria which, interestingly, are justified by the difficulties experienced by more mainstream algorithms in clustering data of the type found in the H-R diagram.)

22. C.T. Zahn, "Graph-theoretical methods for detecting and describing Gestalt clusters", *IEEE Transactions on Computers*, C–**20**, 68–86, 1971.

   (Central reference for the use of the Minimal Spanning Tree for processing point patterns.)

3.4 Software and Sample Implementation

In the first section, following, hierarchical clustering is dealt with. The principal subroutines are HC, which carries out a dissimilarity–based hierarchical clustering; and HCON2, which carries out a hierarchic clustering based on the original data in $O(n^2)$ time. The use of HCASS and HCDEN may follow either: these subroutines determine class assignments and allow a dendrogram to be drawn.

The second section, which follows, deals with partitioning. Following Späth (1985), the essential subroutines are MINDST and EXCH for the minimal distance and exchange methods, respectively.

3.4.1 Program Listing: Hierarchical Clustering

```
C++++++++++++++++++++++++++++++++++++++++++++++++++
C
C    HIERARCHICAL CLUSTERING using (user-specified) criterion.
C
C    Parameters:
C
C    DATA(N,M)         input data matrix,
C    DISS(LEN)         dissimilarities in lower half diagonal
C                      storage; LEN = N.N-1/2,
C    IOPT              clustering criterion to be used,
C    IA, IB, CRIT      history of agglomerations; dimensions
C                      N, first N-1 locations only used,
C    MEMBR, NN, DISNN  vectors of length N, used to store
C                      cluster cardinalities, current nearest
C                      neighbour, and the dissimilarity assoc.
C                      with the latter.
C    FLAG              boolean indicator of agglomerable obj./
C                      clusters.
C
C------------------------------------------------------------
      SUBROUTINE HC(N,M,LEN,IOPT,DATA,IA,IB,CRIT,MEMBR,NN,DISNN,
     X              FLAG,DISS)
      REAL DATA(N,M),MEMBR(N),DISS(LEN)
      INTEGER IA(N),IB(N)
      REAL CRIT(N)
      DIMENSION NN(N),DISNN(N)
      LOGICAL FLAG(N)
      REAL INF
      DATA INF/1.E+20/
C
C    Initializations
C
      DO 10 I=1,N
         MEMBR(I)=1.
         FLAG(I)=.TRUE.
   10 CONTINUE
      NCL=N
C
C    Construct dissimilarity matrix
C
      DO 40 I=1,N-1
         DO 30 J=I+1,N
            IND=IOFFS(N,I,J)
            DISS(IND)=0.
            DO 20 K=1,M
               DISS(IND)=DISS(IND)+(DATA(I,K)-DATA(J,K))**2
   20       CONTINUE
```

3.4. SOFTWARE AND SAMPLE IMPLEMENTATION

```
               IF (IOPT.EQ.1) DISS(IND)=DISS(IND)/2.
C              (Above is done for the case of the min. var. method
C               where merging criteria are defined in terms of
C               variances rather than distances.)
 30         CONTINUE
 40      CONTINUE
C
C  Carry out an agglomeration - first create list of NNs
C
       DO 60 I=1,N-1
          DMIN=INF
          DO 50 J=I+1,N
             IND=IOFFS(N,I,J)
             IF (DISS(IND).GE.DMIN) GOTO 50
                DMIN=DISS(IND)
                JM=J
 50       CONTINUE
          NN(I)=JM
          DISNN(I)=DMIN
 60    CONTINUE
C
 70    CONTINUE
C      Next, determine least diss. using list of NNs
       DMIN=INF
       DO 80 I=1,N-1
          IF (.NOT.FLAG(I)) GOTO 80
          IF (DISNN(I).GE.DMIN) GOTO 80
             DMIN=DISNN(I)
             IM=I
             JM=NN(I)
 80    CONTINUE
       NCL=NCL-1
C
C  This allows an agglomeration to be carried out.
C
       I2=MINO(IM,JM)
       J2=MAXO(IM,JM)
       IA(N-NCL)=I2
       IB(N-NCL)=J2
       CRIT(N-NCL)=DMIN
C
C  Update dissimilarities from new cluster.
C
       FLAG(J2)=.FALSE.
       DMIN=INF
       DO 170 K=1,N
          IF (.NOT.FLAG(K)) GOTO 160
          IF (K.EQ.I2) GOTO 160
          X=MEMBR(I2)+MEMBR(J2)+MEMBR(K)
          IF (I2.LT.K) IND1=IOFFS(N,I2,K)
```

```
              IF (I2.GE.K) IND1=IOFFS(N,K,I2)
              IF (J2.LT.K) IND2=IOFFS(N,J2,K)
              IF (J2.GE.K) IND2=IOFFS(N,K,J2)
              IND3=IOFFS(N,I2,J2)
              XX=DISS(IND3)
C
C   WARD'S MINIMUM VARIANCE METHOD - IOPT=1.
C
              IF (IOPT.NE.1) GOTO 90
                  DISS(IND1)=(MEMBR(I2)+MEMBR(K))*DISS(IND1)+
       X                     (MEMBR(J2)+MEMBR(K))*DISS(IND2)-
       X                      MEMBR(K)*XX
                  DISS(IND1)=DISS(IND1)/X
   90         CONTINUE
C
C   SINGLE LINK METHOD - IOPT=2.
C
              IF (IOPT.NE.2) GOTO 100
                  DISS(IND1)=MIN(DISS(IND1),DISS(IND2))
  100         CONTINUE
C
C   COMPLETE LINK METHOD - IOPT=3.
C
              IF (IOPT.NE.3) GOTO 110
                  DISS(IND1)=MAX(DISS(IND1),DISS(IND2))
  110         CONTINUE
C
C   AVERAGE LINK (OR GROUP AVERAGE) METHOD - IOPT=4.
C
              IF (IOPT.NE.4) GOTO 120
                  DISS(IND1)=(MEMBR(I2)*DISS(IND1)+MEMBR(J2)*DISS(IND2))/
       X                     (MEMBR(I2)+MEMBR(J2))
  120         CONTINUE
C
C   MCQUITTY'S METHOD - IOPT=5.
C
              IF (IOPT.NE.5) GOTO 130
                  DISS(IND1)=0.5*DISS(IND1)+0.5*DISS(IND2)
  130         CONTINUE
C
C   MEDIAN (GOWER'S) METHOD - IOPT=6.
C
              IF (IOPT.NE.6) GOTO 140
                  DISS(IND1)=0.5*DISS(IND1)+0.5*DISS(IND2)-0.25*XX
  140         CONTINUE
C
C   CENTROID METHOD - IOPT=7.
C
```

3.4. SOFTWARE AND SAMPLE IMPLEMENTATION

```
          IF (IOPT.NE.7) GOTO 150
              DISS(IND1)=(MEMBR(I2)*DISS(IND1)+MEMBR(J2)*DISS(IND2)-
     X            MEMBR(I2)*MEMBR(J2)*XX/(MEMBR(I2)+MEMBR(J2)))/
     X            (MEMBR(I2)+MEMBR(J2))
 150      CONTINUE
C
          IF (I2.GT.K) GOTO 160
          IF (DISS(IND1).GE.DMIN) GOTO 160
              DMIN=DISS(IND1)
              JJ=K
 160      CONTINUE
 170  CONTINUE
      MEMBR(I2)=MEMBR(I2)+MEMBR(J2)
      DISNN(I2)=DMIN
      NN(I2)=JJ
C
C  Update list of NNs insofar as this is required.
C
      DO 200 I=1,N-1
          IF (.NOT.FLAG(I)) GOTO 200
          IF (NN(I).EQ.I2) GOTO 180
          IF (NN(I).EQ.J2) GOTO 180
          GOTO 200
 180      CONTINUE
C         (Redetermine NN of I:)
          DMIN=INF
          DO 190 J=I+1,N
              IND=IOFFS(N,I,J)
              IF (.NOT.FLAG(J)) GOTO 190
              IF (I.EQ.J) GOTO 190
              IF (DISS(IND).GE.DMIN) GOTO 190
                  DMIN=DISS(IND)
                  JJ=J
 190      CONTINUE
          NN(I)=JJ
          DISNN(I)=DMIN
 200  CONTINUE
C
C  Repeat previous steps until N-1 agglomerations carried out.
C
      IF (NCL.GT.1) GOTO 70
C
C
      RETURN
      END
C
C
      FUNCTION IOFFS(N,I,J)
C  Map row I and column J of upper half diagonal symmetric matrix
```

```
C     onto vector.
      IOFFS=J+(I-1)*N-(I*(I+1))/2
      RETURN
      END
C++++++++++++++++++++++++++++++++++++++++++++++++++++++++
C
C  HIERARCHICAL CLUSTERING using Minimum Variance Criterion,
C  using the O(N**2) time Nearest Neighbour Chain algorithm.
C
C  Parameters:
C
C  DATA(N,M) :       input data,
C  IA(N), IB(N), CRIT(N) : sequence of agglomerands and
C                    values returned (only locations 1 to N-1
C                    are of interest),
C  MEMBR(N), DISS(N), ICHAIN(N) : used in the routines to store
C                    cluster cardinalities, nearest neighbour
C                    dissimilarities, and the NN-chain.
C  FLAG(N) :         (boolean) used to indicate agglomerable
C                    objects and clusters.
C
C  Reference: Murtagh, Multidimensional Clustering Algorithms,
C             Physica-Verlag, 1985.
C
C---------------------------------------------------------------
      SUBROUTINE HCON2(N,M,DATA,IA,IB,CRIT,MEMBR,DISS,ICHAIN,FLAG)
      REAL    MEMBR(N), DATA(N,M), DISS(N), CRIT(N)
      INTEGER ICHAIN(N), IA(N), IB(N)
      REAL INF
      LOGICAL FLAG(N)
      DATA INF/1.E+25/
C     EQUIVALENCE (ICHAIN(1),IA(1)),(DISS(1),CRIT(1))
C
      DO 150 I=1,N
         MEMBR(I)=1
         FLAG(I)=.TRUE.
  150 CONTINUE
      NCL=N
      I1=1
C
C  Start the NN-chain:
C
  200 LEN=N
      ICHAIN(LEN)=I1
      DISS(LEN)=INF
C
C  Determine NN of object I1
C
  300 FLAG(I1)=.FALSE.
```

3.4. SOFTWARE AND SAMPLE IMPLEMENTATION

```
C
C  Turn off FLAG so that 0 diss. of I1 with self not obtained.
C
      D=DISS(LEN)
      IF (LEN.LT.N) I2=ICHAIN(LEN+1)
C
C  For identical diss.'s, above ensures that RNN will be found.
C
      CALL DETNN(DATA,FLAG,MEMBR,N,M,I1,I2,D)
      FLAG(I1)=.TRUE.
C
C  If LEN = 1 place obj. I2 as second obj. in NN-chain.
C
      IF (LEN.LT.N) GOTO 350
      LEN=LEN-1
      IF (LEN.LT.N-NCL) GOTO 700
      ICHAIN(LEN)=I2
      DISS(LEN)=D
      GOTO 500
C
C  If LEN < N distinguish between having RNN & continuing NN-chain.
C
  350 CONTINUE
      IF (I2.NE.ICHAIN(LEN+1)) GOTO 400
C
C  Have RNN.
C
      NCL=NCL-1
      CALL AGGLOM(I1,I2,D,DATA,MEMBR,FLAG,IA,IB,CRIT,NCL,N,M)
      LEN=LEN+2
      GOTO 500
  400 CONTINUE
C
C  Grow extra link on NN-chain.
C
      IDUM=ICHAIN(LEN+1)
      FLAG(IDUM)=.FALSE.
      LEN=LEN-1
      IF (LEN.LE.N-NCL) GOTO 700
      ICHAIN(LEN)=I2
      DISS(LEN)=D
      GOTO 500
C
C  Select obj. for continuing to grow (or restarting) NN-chain.
C
  500 CONTINUE
      IF (NCL.EQ.1) GOTO 600
      IF (LEN.EQ.N+1) GOTO 550
      I1=ICHAIN(LEN)
```

```
          FLAG(I1)=.TRUE.
          IDUM=ICHAIN(LEN+1)
          IF (LEN.LT.N) FLAG(IDUM)=.TRUE.
C
C   Reestablish agglomerability of objects in NN-chain.
C
          GOTO 300
   550    CALL NEXT(FLAG,I1,N)
          GOTO 200
C
   600    CONTINUE
          RETURN
   700    WRITE(6,750)
   750    FORMAT
         X(' ERROR IN NN-CHAIN ROUTINE - INSUFFICIENT CHAIN SPACE'/)
          STOP
          END
C-------------------------------------------------------------
          SUBROUTINE DETNN(DATA,FLAG,MEM,N,M,I1,I2,D)
C
C   Determine a nearest neighbour.
C
          REAL DATA(N,M),MEM(N)
          LOGICAL FLAG(N)
C
          DO 200 I=1,N
              IF (.NOT.FLAG(I)) GOTO 200
              DISS=0.
              DO 100 J=1,M
   100        DISS=DISS+(DATA(I1,J)-DATA(I,J))*(DATA(I1,J)-DATA(I,J))
              DISS=DISS*MEM(I)*MEM(I1)/(MEM(I1)+MEM(I))
              IF (DISS.GE.D) GOTO 200
                  D=DISS
                  I2=I
   200    CONTINUE
C
          RETURN
          END
C-------------------------------------------------------------
          SUBROUTINE AGGLOM(I1,I2,D,DATA,MEM,FLAG,IA,IB,CRIT,NCL,N,M)
C
C   Carry out an agglomeration.
C
          REAL MEM(N),DATA(N,M),CRIT(N)
          INTEGER IA(N),IB(N)
          LOGICAL FLAG(N)
          INTEGER O1,O2,LB,UB
C
C
```

3.4. SOFTWARE AND SAMPLE IMPLEMENTATION

```
      O1=MIN0(I1,I2)
      O2=MAX0(I1,I2)
      DO 100 J=1,M
         DATA(O1,J)=( MEM(O1)*DATA(O1,J)+MEM(O2)*DATA(O2,J) )
     X            / (MEM(O1)+MEM(O2))
         DATA(O2,J)=DATA(O1,J)
  100 CONTINUE
      NAGGL=N-NCL
      MEM(O1)=MEM(O1)+MEM(O2)
      FLAG(O2)=.FALSE.
C
C  Keep sorted list of criterion values: find first where
C  new criterion value fits.
C
      I=NAGGL-1
  120 IF (D.GE.CRIT(I)) GOTO 140
      I=I-1
      IF (I.GE.1) GOTO 120
C
C  Arriving here must mean that D > all crit. values found so far.
C
      I=0
  140 CONTINUE
C
C  Now, shift rightwards from I+1 to AGGL-1 to make room for
C  new criterion value.
C
      LB=I+1
      UB=NAGGL-1
      IF (LB.GT.UB) GOTO 180
      J=UB
  160 J1=J+1
      IA(J1)=IA(J)
      IB(J1)=IB(J)
      CRIT(J1)=CRIT(J)
      J=J-1
      IF (J.GE.LB) GOTO 160
  180 CONTINUE
      IA(LB)=O1
      IB(LB)=O2
      CRIT(LB)=D
C
      RETURN
      END
C-----------------------------------------------------------
      SUBROUTINE NEXT(FLAG,I1,N)
C
C  Determine next agglomerable object/cluster.
C
```

```
      LOGICAL FLAG(N)
C
      NXT=I1+1
      IF (NXT.GT.N) GOTO 150
      DO 100 I=NXT,N
          IF (FLAG(I)) GOTO 500
 100  CONTINUE
 150  DO 200 I=1,I1
          IF (FLAG(I)) GOTO 500
 200  CONTINUE
C
      STOP
C
 500  I1=I
C
      RETURN
      END
C++++++++++++++++++++++++++++++++++++++++++++++++++++
C
C  Given a HIERARCHIC CLUSTERING, described as a sequence of
C  agglomerations, derive the assignments into clusters for the
C  top LEV-1 levels of the hierarchy.
C  Prepare also the required data for representing the
C  dendrogram of this top part of the hierarchy.
C
C  Parameters:
C
C  IA, IB, CRIT: vectors of dimension N defining the agglomer-
C                ations.
C  LEV:          number of clusters in largest partition.
C  HVALS:        vector of dim. LEV, used internally only.
C  ICLASS:       array of cluster assignments; dim. N by LEV.
C  IORDER, CRITVL, HEIGHT: vectors describing the dendrogram,
C                all of dim. LEV.
C
C  N should be greater than LEV.
C
C-------------------------------------------------------------
      SUBROUTINE HCASS(N,IA,IB,CRIT,LEV,ICLASS,HVALS,IORDER,
     X           CRITVL,HEIGHT)
      INTEGER IA(N),IB(N),ICLASS(N,LEV),HVALS(LEV),IORDER(LEV),
     X           HEIGHT(LEV)
      REAL CRIT(N),CRITVL(LEV)
C
C  Pick out the clusters which the N objects belong to,
C  at levels N-2, N-3, ... N-LEV+1 of the hierarchy.
C  The clusters are identified by the lowest seq. no. of
C  their members.
C  There are 2, 3, ... LEV clusters, respectively, for the
```

3.4. SOFTWARE AND SAMPLE IMPLEMENTATION

```
C     above levels of the hierarchy.
C
      HVALS(1)=1
      HVALS(2)=IB(N-1)
      LOC=3
      DO 59 I=N-2,N-LEV,-1
         DO 52 J=1,LOC-1
            IF (IA(I).EQ.HVALS(J)) GOTO 54
  52     CONTINUE
         HVALS(LOC)=IA(I)
         LOC=LOC+1
  54     CONTINUE
         DO 56 J=1,LOC-1
            IF (IB(I).EQ.HVALS(J)) GOTO 58
  56     CONTINUE
         HVALS(LOC)=IB(I)
         LOC=LOC+1
  58     CONTINUE
  59  CONTINUE
C
      DO 400 LEVEL=N-LEV,N-2
         DO 200 I=1,N
            ICL=I
            DO 100 ILEV=1,LEVEL
 100        IF (IB(ILEV).EQ.ICL) ICL=IA(ILEV)
            NCL=N-LEVEL
            ICLASS(I,NCL-1)=ICL
 200     CONTINUE
 400  CONTINUE
C
      DO 120 I=1,N
      DO 120 J=1,LEV-1
      DO 110 K=2,LEV
      IF (ICLASS(I,J).NE.HVALS(K)) GOTO 110
         ICLASS(I,J)=K
         GOTO 120
 110  CONTINUE
 120  CONTINUE
C
      WRITE (6,450)
 450  FORMAT(4X,' SEQ NOS 2CL 3CL 4CL 5CL 6CL 7CL 8CL 9CL')
      WRITE (6,470)
 470  FORMAT(4X,' ------- --- --- --- --- --- --- --- ----')
      DO 500 I=1,N
      WRITE (6,600) I,(ICLASS(I,J),J=1,8)
 600  FORMAT(I11,8I4)
 500  CONTINUE
C
C     Determine an ordering of the LEV clusters (at level LEV-1)
```

```
C     for later representation of the dendrogram.
C     These are stored in IORDER.
C     Determine the associated ordering of the criterion values
C     for the vertical lines in the dendrogram.
C     The ordinal values of these criterion values may be used in
C     preference, and these are stored in HEIGHT.
C     Finally, note that the LEV clusters are renamed so that they
C     have seq. nos. 1 to LEV.
C
      IORDER(1)=IA(N-1)
      IORDER(2)=IB(N-1)
      CRITVL(1)=0.0
      CRITVL(2)=CRIT(N-1)
      HEIGHT(1)=LEV
      HEIGHT(2)=LEV-1
      LOC=2
      DO 700 I=N-2,N-LEV+1,-1
         DO 650 J=1,LOC
            IF (IA(I).EQ.IORDER(J)) THEN
C              Shift rightwards and insert IB(I) beside IORDER(J):
               DO 630 K=LOC+1,J+1,-1
                  IORDER(K)=IORDER(K-1)
                  CRITVL(K)=CRITVL(K-1)
                  HEIGHT(K)=HEIGHT(K-1)
  630          CONTINUE
               IORDER(J+1)=IB(I)
               CRITVL(J+1)=CRIT(I)
               HEIGHT(J+1)=I-(N-LEV)
               LOC=LOC+1
            ENDIF
  650    CONTINUE
  700 CONTINUE
      DO 705 I=1,LEV
         DO 703 J=1,LEV
            IF (HVALS(I).EQ.IORDER(J)) THEN
               IORDER(J)=I
               GOTO 705
            ENDIF
  703    CONTINUE
  705 CONTINUE
C
      RETURN
      END
C+++++++++++++++++++++++++++++++++++++++++++++++
C
C     Construct a DENDROGRAM of the top 8 levels of
C     a HIERARCHIC CLUSTERING.
C
C     Parameters:
```

3.4. SOFTWARE AND SAMPLE IMPLEMENTATION

```
C
C   IORDER, HEIGHT, CRITVL: vectors of length LEV
C           defining the dendrogram.
C           These are: the ordering of objects
C           along the bottom of the dendrogram
C           (IORDER); the height of the vertical
C           above each object, in ordinal values
C           (HEIGHT); and in real values (CRITVL).
C
C   NOTE: these vectors MUST have been set up with
C         LEV = 9 in the prior call to routine
C         HCASS.
C
C-----------------------------------------------
      SUBROUTINE HCDEN(LEV,IORDER,HEIGHT,CRITVL)
      CHARACTER*80 LINE
      INTEGER IORDER(LEV),HEIGHT(LEV)
      REAL CRITVL(LEV)
      INTEGER OUT(27,27)
      INTEGER UP,ACROSS,BLANK
      DATA UP,ACROSS,BLANK/'|','-',' '/
C
C
      DO 10 I=1,27
        DO 10 J=1,27
          OUT(I,J)=BLANK
 10   CONTINUE
C
C
      DO 50 I=3,27,3
        I2=I/3
C
        J2=28-3*HEIGHT(I2)
        J = 27
 20     CONTINUE
          OUT(J,I)=UP
          J = J-1
        IF (J.GE.J2) GOTO 20
C
        K = I
 30     CONTINUE
          I3=INT((K+2)/3)
          IF ( (28-HEIGHT(I3)*3).LT.J2) GOTO 40
          OUT(J2,K)=ACROSS
          K = K-1
        IF (K.GE.3) GOTO 30
 40     CONTINUE
C
 50   CONTINUE
```

```
      C
      C
            IC=3
            DO 90 I=1,27
            IF (I.NE.IC+1) GOTO 80
                     IDUM=IC/3
                     IDUM=9-IDUM
                     DO 60 L=1,9
                        IF (HEIGHT(L).EQ.IDUM) GOTO 70
   60                CONTINUE
   70                IDUM=L
                     WRITE(6,200) CRITVL(IDUM),(OUT(I,J),J=1,27)
                     IC=IC+3
                     GOTO 90
   80                CONTINUE
                     LINE =
                     WRITE(6,210) (OUT(I,J),J=1,27)
   90       CONTINUE
            WRITE(6,250)
            WRITE(6,220)(IORDER(J),J=1,9)
            WRITE(6,250)
            WRITE(6,230)
            WRITE(6,240)
  200       FORMAT(1H ,8X,F12.2,4X,27A1)
  210       FORMAT(1H ,24X,27A1)
  220       FORMAT(1H ,24X,9I3)
  230       FORMAT(1H ,13X,'CRITERION      CLUSTERS 1 TO 9')
  240       FORMAT(1H ,13X,'VALUES.    (TOP 8 LEVELS OF HIERARCHY).')
  250       FORMAT(/)
      C
      C
            RETURN
            END
```

3.4.2 Program Listing: Partitioning

```
C++++++++++++++++++++++++++++++++++++++++++++++++++
C
C  Generate a random partition.
C
C  Parameters:
C
C  MEMGP(N)        Group memberships,
C  NG              number of groups,
C  ISEED           seed for random number generator.
C
C  Note: random number generator is machine-dependent.
C
C------------------------------------------------------------
       SUBROUTINE RANDP(N,NG,MEMGP,ISEED)
       DIMENSION MEMGP(N)
C
       DO 100 I = 1, N
          MEMGP(I) = 1
  100  CONTINUE
C
       IF (NG.LE.1.OR.N.LE.1) GOTO 500
       X = 1.0/FLOAT(NG)
       DO 400 I = 1, N
          VAL = RAN(ISEED)
          BNDRY = X
          ICL = 1
  200     IF (ICL.EQ.NG) GOTO 300
          IF (VAL.LT.BNDRY) GOTO 300
          BNDRY = BNDRY + X
          ICL = ICL + 1
          GOTO 200
  300     MEMGP(I) = ICL
  400  CONTINUE
C
  500  CONTINUE
       RETURN
       END
C++++++++++++++++++++++++++++++++++++++++++++++++++
C
C  Optimise the variances of a set of groups, by assigning
C  the objects in groups such that they are minimally distant
C  from group centres.
C
C  Parameters:
C
C  N, M, NG        Numbers of rows, columns, groups,
C  A(N,M)          initial data,
```

```
C     MEMGP(N)           group memberships,
C     NGPO               minimum acceptable group cardinality,
C     NUMGP(NG)          cardinalities of groups,
C     GPCEN(NG,M)        group centres,
C     COMP(NG)           compactness values for the groups,
C     CTOT               sum of these compactnesses,
C     IERR               error indicator (should be zero).
C
C     IERR = 1: invalid group number (<1 or >NG), - is number of
C     groups correctly specified?  IERR = 2: a group has < minimum
C     allowed number of members, - reduce the number of groups and
C     try again.
C
C     Reference: Spaeth, 1985.
C
C-----------------------------------------------------------------
      SUBROUTINE MINDST(A,N,M,MEMGP,NGPO,NUMGP,GPCEN,NG,
     X                  COMP,CTOT,ITER,IERR)
      DIMENSION         A(N,M), MEMGP(N), NUMGP(NG), GPCEN(NG,M),
     X                  COMP(NG)
C
      BIG = 1.0E+30
      ONE = 0.999

      CMAX = BIG
      ITER = 0
  100 ITER = ITER + 1
      IF (ITER.GT.15) GOTO 500
      CALL GMEANS(A,N,M,MEMGP,NGPO,NUMGP,GPCEN,NG,IERR)
      CALL COMPCT(A,N,M,NG,MEMGP,GPCEN,COMP,CTOT)
      IF (IERR.NE.0) GOTO 500
      IF (NG.LE.1) GOTO 500
      IF (CTOT.GE.CMAX) GOTO 500
      CMAX = CTOT*ONE
      DO 400 I = 1, N
         X = BIG
         DO 300 K = 1, NG
            Y = 0.0
            DO 200 J = 1, M
               DIFF = GPCEN(K,J) - A(I,J)
               Y = Y + DIFF*DIFF
  200       CONTINUE
            IF (Y.GE.X) GOTO 300
            X = Y
            ICL = K
  300    CONTINUE
         MEMGP(I) = ICL
  400 CONTINUE
      GOTO 100
C
```

3.4. SOFTWARE AND SAMPLE IMPLEMENTATION

```
  500 RETURN
      END
C+++++++++++++++++++++++++++++++++++++++++++++++++++++++
C
C  Optimise the variances of a set of groups, by exchanging
C  the objects between groups such that they are minimally
C  distant from group centres.
C
C  Parameters:
C
C  N, M, NG         Numbers of rows, columns, groups,
C  A(N,M)           initial data,
C  MEMGP(N)         group memberships,
C  NGPO             minimum acceptable group cardinality,
C  NUMGP(NG)        cardinalities of groups,
C  GPCEN(NG,M)      group centres,
C  COMP(NG)         compactness values for the groups,
C  CTOT             sum of these compactnesses,
C  IERR             error indicator (should be zero).
C
C  IERR = 1: invalid group number (<1 or >NG), - is number of
C  groups correctly specified?  IERR = 2: a group has < minimum
C  allowed number of members, - reduce the number of groups and
C  try again.
C
C  Reference: Spaeth, 1985.
C
C-----------------------------------------------------------------
      SUBROUTINE EXCH(A,N,M,MEMGP,NGPO,NUMGP,GPCEN,NG,
     X                COMP,CTOT,ITER,IERR)
      DIMENSION       A(N,M), MEMGP(N), NUMGP(NG), GPCEN(NG,M),
     X                COMP(NG)
C
      BIG = 1.0E+30
      ONE = 0.999
      CALL GMEANS(A,N,M,MEMGP,NGPO,NUMGP,GPCEN,NG,IERR)
      CALL COMPCT(A,N,M,NG,MEMGP,GPCEN,COMP,CTOT)
      IF (IERR.NE.0) GOTO 800
      IF (NG.LE.1) GOTO 800
      ITER = 0
      I = 0
      IS = 0
  100 IS = IS + 1
      IF (IS.GT.N) GOTO 800
  200 I = I + 1
      IF (I.LE.N) GOTO 300
      ITER = ITER + 1
      IF (ITER.GT.15) GOTO 800
      I = 1
```

```
      300   ICL = MEMGP(I)
            NUM = NUMGP(ICL)
            IF (NUM.LE.NGPO) GOTO 100
            V = NUM
            EQ = BIG
            DO 600 K = 1, NG
               X = 0.0
               DO 400 J = 1, M
                  DIFF = GPCEN(K,J) - A(I,J)
                  X = X + DIFF*DIFF
      400      CONTINUE
               IF (K.NE.ICL) GOTO 500
               FRAC1 = V/(V-1.0)
               EP = X*FRAC1
               GOTO 600
      500      FRAC2 = NUMGP(K)
               FRAC = FRAC2/(FRAC2+1.0)
               EK = FRAC*X
               IF (EK.GE.EQ) GOTO 600
               EQ = EK
               IQ = K
               W = FRAC2
      600   CONTINUE
            IF (EQ.GE.EP*ONE) GOTO 100
            IS = 0
            COMP(ICL) = COMP(ICL) - EP
            COMP(IQ) = COMP(IQ) + EQ
            CTOT = CTOT - EP + EQ
            FRAC1 = 1.0/(V-1.0)
            FRAC2 = 1.0/(W+1.0)
            DO 700 J = 1, M
               VAL = A(I,J)
               GPCEN(ICL,J) = (V*GPCEN(ICL,J)-VAL)*FRAC1
               GPCEN(IQ,J) = (W*GPCEN(IQ,J)+VAL)*FRAC2
      700   CONTINUE
            MEMGP(I) = IQ
            NUMGP(ICL) = NUM - 1
            NUMGP(IQ) = NUMGP(IQ) + 1
            GOTO 200
C
      800   CONTINUE
            RETURN
            END
C++++++++++++++++++++++++++++++++++++++++++++++++++++++++
C
C   Standardize to zero mean and unit standard deviation.
C
C   Parameters:
C
```

3.4. SOFTWARE AND SAMPLE IMPLEMENTATION

```
C   N, M, NG          Numbers of rows, columns, groups,
C   A(N,M)            initial data, replaced by standardized values.
C
C-----------------------------------------------------------
        SUBROUTINE STND(A,N,M)
        DIMENSION  A(N,M)
C
        DO 500 J = 1, M
           X = 0.0
           DO 100 I = 1, N
              X = X + A(I,J)
  100      CONTINUE
           XBAR = X/FLOAT(N)
           X = 0.0
           DO 200 I = 1, N
              DIFF = A(I,J) - XBAR
              X = X + DIFF*DIFF
  200      CONTINUE
           IF (X.LE.0.0) X = 1.0
           X = 1.0/SQRT(X)
           DO 300 I = 1, N
              A(I,J) = X*(A(I,J)-XBAR)
  300      CONTINUE
  500   CONTINUE
C
        RETURN
        END
C++++++++++++++++++++++++++++++++++++++++++++++++++++++++++
C
C   Determine means of the groups.
C
C   Parameters:
C
C   N, M, NG          Numbers of rows, columns, groups,
C   A(N,M)            initial data,
C   MEMGP(N)          group memberships,
C   NGPO              minimum acceptable group cardinality,
C   NUMGP(NG)         cardinalities of groups,
C   GPCEN(NG,M)       group centres,
C   IERR              error indicator (should be zero).
C
C-----------------------------------------------------------
        SUBROUTINE GMEANS(A,N,M,MEMGP,NGPO,NUMGP,GPCEN,NG,IERR)
        DIMENSION    A(N,M), MEMGP(N), NUMGP(NG), GPCEN(NG,M)
C
        DO 200 K = 1, NG
           NUMGP(K) = 0
           DO 100 J = 1, M
              GPCEN(K,J) = 0.0
```

```
      100       CONTINUE
      200       CONTINUE
C
            DO 500 I = 1, N
               ICL = MEMGP(I)
               IF (ICL.GE.1.AND.ICL.LE.NG) GOTO 300
                  IERR = 1
                  RETURN
      300       CONTINUE
               NUMGP(ICL) = NUMGP(ICL) + 1
               DO 400 J = 1, M
                  GPCEN(ICL,J) = GPCEN(ICL,J) + A(I,J)
      400       CONTINUE
      500    CONTINUE
C
            DO 800 K = 1, NG
               NUM = NUMGP(K)
               IF (NUM.GE.NGPO) GOTO 600
                  IERR = 2
                  RETURN
      600       CONTINUE
               X = 1.0/FLOAT(NUM)
               DO 700 J = 1, M
                  GPCEN(K,J) = GPCEN(K,J)*X
      700       CONTINUE
      800    CONTINUE
C

            RETURN
            END
C++++++++++++++++++++++++++++++++++++++++++++++++++++++++
C
C  Determine compactness of the groups (i.e. their variances).
C
C  Parameters:
C
C  N, M, NG         Numbers of rows, columns, groups,
C  A(N,M)           initial data,
C  MEMGP(N)         group memberships,
C  GPCEN(NG,M)      group centres,
C  COMP(NG)         variances of groups (output),
C  CTOT             sum of these variances.
C
C----------------------------------------------------------------
            SUBROUTINE COMPCT(A,N,M,NG,MEMGP,GPCEN,COMP,CTOT)
            DIMENSION     A(N,M), MEMGP(N), GPCEN(NG,M), COMP(NG)
C
            CTOT = 0.0
            DO 100 K = 1, NG
               COMP(K) = 0.0
```

3.4. SOFTWARE AND SAMPLE IMPLEMENTATION

```
  100     CONTINUE
C
        DO 300 I = 1, N
           ICL = MEMGP(I)
           X = 0.0
           DO 200 J = 1, M
              DIFF = GPCEN(ICL,J) - A(I,J)
              X = X + DIFF*DIFF
  200      CONTINUE
           COMP(ICL) = COMP(ICL) + X
           CTOT = CTOT + X
  300   CONTINUE
C
        RETURN
        END
```

3.4.3 Input Data

The input data is identical to that described in the Principal Components Analysis chapter.

3.4.4 Sample Output

For the minimum variance criterion, we find the following upper dendrogram result. Note that class numbers are purely sequence numbers.

SEQ NOS	2CL	3CL	4CL	5CL	6CL	7CL	8CL	9CL
1	1	1	1	1	1	1	1	1
2	1	1	1	1	1	1	1	1
3	1	1	1	1	1	1	1	1
4	1	1	1	1	1	1	1	1
5	1	1	1	1	1	1	1	1
6	1	1	1	1	1	1	1	1
7	1	1	1	1	1	7	7	7
8	1	1	1	1	1	7	7	7
9	1	1	1	1	1	7	7	9
10	1	1	1	1	1	7	7	9
11	1	1	4	4	4	4	4	4
12	1	1	4	4	4	4	4	4
13	1	1	4	4	4	4	8	8
14	1	1	4	4	4	4	8	8
15	1	1	4	5	5	5	5	5
16	2	2	2	2	2	2	2	2
17	2	2	2	2	6	6	6	6
18	2	3	3	3	3	3	3	3

3.4. SOFTWARE AND SAMPLE IMPLEMENTATION

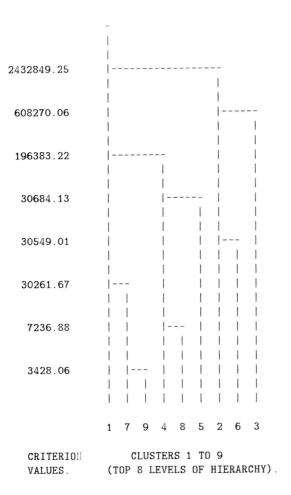

CRITERION   CLUSTERS 1 TO 9
VALUES.   (TOP 8 LEVELS OF HIERARCHY).

The following dendrogram output is obtained for the single link method.

SEQ NOS	2CL	3CL	4CL	5CL	6CL	7CL	8CL	9CL
1	1	1	1	1	1	1	1	1
2	1	1	1	1	1	1	1	1
3	1	1	1	1	1	1	1	1

```
 4   1   1   1   1   1   1   1   1
 5   1   1   1   1   1   1   1   1
 6   1   1   1   1   1   1   1   1
 7   1   1   1   1   1   1   1   9
 8   1   1   1   1   1   1   1   9
 9   1   1   1   1   1   1   1   9
10   1   1   1   1   1   1   1   9
11   1   1   1   1   1   7   7   7
12   1   1   1   1   1   7   7   7
13   1   1   1   1   1   7   8   8
14   1   1   1   1   6   6   6   6
15   1   1   1   5   5   5   5   5
16   1   3   3   3   3   3   3   3
17   1   3   4   4   4   4   4   4
18   2   2   2   2   2   2   2   2
```

```
691853.06   |------------------------
                                    |
                                    |
 64923.62   |-----------------       |
                           |         |
                           |         |
 61098.02   |              |   |---  |
                           |   |  |  |
                           |   |  |  |
 20083.52   |--------------    |  |  |
                           |   |  |  |
                           |   |  |  |
  2809.86   |----------    |   |  |  |
                      |    |   |  |  |
                      |    |   |  |  |
  2565.14   |------   |    |   |  |  |
                  |   |    |   |  |  |
                  |   |    |   |  |  |
  2539.18   |     |---  |  |   |  |  |
            |     | |   |  |   |  |  |
            |     | |   |  |   |  |  |
  1519.50   |---  | |   |  |   |  |  |
            | |   | |   |  |   |  |  |
            | |   | |   |  |   |  |  |

            1  9  7 8   6  5   3  4  2
```

CRITERION   CLUSTERS 1 TO 9
VALUES.   (TOP 8 LEVELS OF HIERARCHY).

3.4. SOFTWARE AND SAMPLE IMPLEMENTATION

For the minimum distance partitioning method, with two classes requested, a solution was found with sum of variances equal to 914415.4375, after 5 iterations. This solution assigned the first 15 objects to class 1 and the final 3 objects to class 2.

For the exchange method, and again with two classes requested, the sum of variances was 906848.4375 after 2 iterations. For this solution, only the final two objects comprised class 2.

For three classes, and the exchange method, the sum of variances was 297641.7813 after 1 iteration. The classes were: object 18; objects 15, 16, 17; and all remaining objects.

Chapter 4

Discriminant Analysis

4.1 The Problem

Discriminant Analysis may be used for two objectives: either we want to *assess* the adequacy of classification, given the group memberships of the objects under study; or we wish to *assign* objects to one of a number of (known) groups of objects. Discriminant Analysis may thus have a descriptive or a predictive objective.

In both cases, some group assignments must be known before carrying out the Discriminant Analysis. Such group assignments, or labelling, may be arrived at in any way. Hence Discriminant Analysis can be employed as a useful complement to Cluster Analysis (in order to judge the results of the latter) or Principal Components Analysis. Alternatively, in star–galaxy separation, for instance, using digitised images, the analyst may define group (stars, galaxies) membership visually for a conveniently small *training set* or *design set*.

The basic ideas used in Discriminant Analysis are quite simple. For assessment of a classification, we attempt to optimally locate a curve — often a straight line — between the classes. For assignment, we basically look for the class which is closest to the new object that we wish to classify. As with other multivariate techniques, a variety of Discriminant Analysis methods are available, based on the precise meaning to be given to such terms as "optimal" and "closeness".

A brief "appetizer" of the range of Discriminant Analysis techniques that are widely used is as follows. It might be assumed for instance that the groups to be distinguished are of Gaussian distribution; or a very flexible separation surface might be used, such as would be entailed when assigning a new sample to the cluster which the k closest neighbours of the new sample belonged to (for some small constant k). Linear discrimination (or, in higher dimensional spaces, hyperplane separation) is both mathematically tractable, and may fulfil our objective of optimally separating classes. If the clusters are intertwined (see Figure 4.1b), linear separation will perform badly; on the other hand in Figure 4.1a, there are only

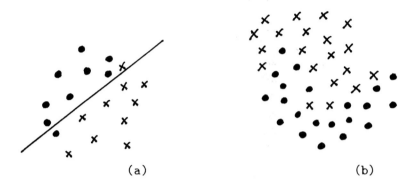

Figure 4.1: Two sets of two groups. Linear separation performs adequately in (a) but non-linear separation is more appropriate in (b).

two points misclassified. Hyperplane separation performs well when the classes are reasonably separated, and assumes no distributional or other properties on the part of the data.

In the following, we will adopt two distinct frames of reference for the description of discriminant methods. On the one hand, we will look at a geometrical framework, which nicely complements the type of perspective employed in PCA. On the other hand, we will overview discriminant methods which make use of a probabilistic perspective. Interestingly, one well known method (Linear Discriminant Analysis) may be specified using *either* mathematical framework.

The fact that Discriminant Analysis is often embraced under the term "classification" means that care should be taken in distinguishing between Cluster and Discriminant Analyses. The former is *unsupervised classification*: no prior knowledge of the group memberships is usually required. Discriminant Analysis, when used for confirmatory or assessment purposes, is *supervised classification* and has been referred to as "learning with a teacher". Because of possible confusion in what are usually quite distinct problems and ranges of methods, we have generally eschewed the use of the term "classification".

4.2 Mathematical Description

4.2.1 Multiple Discriminant Analysis

Multiple Discriminant Analysis (MDA) is also termed Discriminant Factor Analysis and Canonical Discriminant Analysis. It adopts a similar perspective to PCA: the rows of the data matrix to be examined constitute points in a multidimensional space, as also do the group mean vectors. Discriminating axes are determined in this space, in such a way that optimal separation of the predefined groups is attained. As with PCA, the problem becomes mathematically the eigenreduction of a real, symmetric matrix.

Consider the set of objects, $i \in I$; they are characterised by a finite set of parameters, $j \in J$. The vectors associated with the objects are given as the row vectors of the matrix $X = \{x_{ij}\}$. The grand mean of all vectors is given by

$$g_j = \frac{1}{n} \sum_{i \in I} x_{ij}$$

(where n is the cardinality of I). Let y_j be the jth coordinate of the mean of group y; i.e.

$$y_j = \frac{1}{n_y} \sum_{i \in y} x_{ij}$$

(where n_y is the cardinality of group y). Finally, we consider the case of mutually disjoint groups, y, whose union gives I. Let Y be the set of these groups, and let n_Y be the number of groups considered. Evidently, $n_Y \leq n$.

We now define the following three variance–covariance matrices. T (of jk^{th} term, t_{jk}) is the total covariance matrix; W is the within classes covariance matrix; and B is the between classes covariance matrix:

$\mathbf{T} : t_{jk} = \frac{1}{n} \sum_{i \in I} (x_{ij} - g_j)(x_{ik} - g_k)$

$\mathbf{W} : w_{jk} = \frac{1}{n} \sum_{y \in Y} \sum_{i \in y} (x_{ij} - y_j)(x_{ik} - y_k)$

$\mathbf{B} : b_{jk} = \sum_{y \in Y} \frac{n_y}{n} (y_j - g_j)(y_k - g_k)$.

The three matrices, T, W and B, are of dimensions $m \times m$ where m is the number of attributes (i.e. the vectors considered, their grand mean, and the group means are located in $\mathbb{R}^m$).

Generalizing Huyghen's Theorem in classical mechanics, we have that $T = W + B$. This is proved as follows. We have, for the $(j, k)^{th}$ terms of these matrices:

$$\frac{1}{n} \sum_{i \in I} (x_{ij} - g_j)(x_{ik} - g_k)$$

$$= \frac{1}{n} \sum_{y \in Y} \sum_{i \in y} (x_{ij} - y_j)(x_{ik} - y_k) + \sum_{y \in Y} \frac{n_y}{n}(y_j - g_j)(y_k - g_k).$$

Rewriting the first term on the right hand side of the equation as

$$\frac{1}{n} \sum_{y \in Y} \sum_{i \in y} ((x_{ij} - g_j) - (y_j - g_j))((x_{ik} - g_k) - (y_k - g_k))$$

and expanding gives the required result.

The variance of the points in $\mathbb{R}^m$ along any given axis $\mathbf{u}$ is given by $\mathbf{u}'T\mathbf{u}$ (cf. the analogous situation in Principal Components Analysis). The variance of class means along this axis is $\mathbf{u}'B\mathbf{u}$. Finally, the total of the variances within classes along this axis is $\mathbf{u}'W\mathbf{u}$. Since $T = W + B$, we have that

$$\mathbf{u}'T\mathbf{u} = \mathbf{u}'B\mathbf{u} + \mathbf{u}'W\mathbf{u}.$$

The optimal discrimination of the given groups is carried out as follows. We choose axis $\mathbf{u}$ to maximize the spread of class means, while restraining the compactness of the classes, i.e.

$$\max \frac{\mathbf{u}'B\mathbf{u}}{\mathbf{u}'W\mathbf{u}}.$$

This maximization problem is the same as

$$\min \frac{\mathbf{u}'W\mathbf{u}}{\mathbf{u}'B\mathbf{u}} = \min \frac{\mathbf{u}'W\mathbf{u}}{\mathbf{u}'B\mathbf{u}} + 1 = \min \frac{\mathbf{u}'W\mathbf{u} + \mathbf{u}'B\mathbf{u}}{\mathbf{u}'B\mathbf{u}}$$

$$= \min \frac{\mathbf{u}'T\mathbf{u}}{\mathbf{u}'B\mathbf{u}} = \max \frac{\mathbf{u}'B\mathbf{u}}{\mathbf{u}'T\mathbf{u}}.$$

As in PCA (refer to Chapter 2), we use λ as a Lagrangian multiplier, and differentiate the expression $\mathbf{u}'B\mathbf{u} - \lambda(\mathbf{u}'T\mathbf{u})$ with respect to $\mathbf{u}$. This yields $\mathbf{u}$ as the eigenvector of $T^{-1}B$ associated with the largest eigenvalue, λ. Eigenvectors associated with successively large eigenvalues define discriminating factors or axes which are orthogonal to those previously obtained. We may therefore say that MDA is the PCA of a set of centred vectors (the group means) in the T^{-1}-metric.

A difficulty has not been mentioned in the foregoing: the matrix product, $T^{-1}B$ is not necessarily symmetric, and so presents a problem for diagonalization. This difficulty is circumvented as follows. We have that $B\mathbf{u} = \lambda T\mathbf{u}$. Writing B as the product of its square roots CC' (which we can always do because of the fact that B is necessarily positive definite and symmetric) gives: $CC'\mathbf{u} = \lambda T\mathbf{u}$. Next, define a new vector $\mathbf{a}$ as follows: $\mathbf{u} = T^{-1}C\mathbf{a}$. This gives:

$$CC'T^{-1}C\mathbf{a} = \lambda TT^{-1}C\mathbf{a}$$

4.2. MATHEMATICAL DESCRIPTION

$$\Rightarrow C(C'T^{-1}C)\mathbf{a} = \lambda C\mathbf{a}$$

$$\Rightarrow (C'T^{-1}C)\mathbf{a} = \lambda \mathbf{a}.$$

We now have an eigenvalue equation, which has a matrix which is necessarily real and symmetric. This is solved for $\mathbf{a}$, and substituted back to yield $\mathbf{u}$.

Since the largest eigenvalue is

$$\frac{\mathbf{u}'B\mathbf{u}}{\mathbf{u}'T\mathbf{u}} = \frac{\mathbf{u}'T\mathbf{u} - \mathbf{u}'W\mathbf{u}}{\mathbf{u}'T\mathbf{u}},$$

it is seen that the right side here, and hence all eigenvalues, are necessarily positive and less than 1.

The eigenvalues represent the discriminating power of the associated eigenvectors. Unlike in PCA, the percentage variance explained by a factor has no sense in MDA, since the sum of eigenvalues has no meaning.

The n_Y classes lie in a space of dimension at most $n_Y - 1$. This will be the number of discriminant axes or factors obtainable in the most common practical case when $n > m > n_Y$.

4.2.2 Linear Discriminant Analysis

In this section, we will examine the 2-group case of MDA, and focus on the assignment problem. Discriminant Analysis may be used for assigning a new object, as well as for confirming the separation between given groups. The distance, in this new T^{-1}-metric, between some new vector $\mathbf{a}$ and the barycentre (or centre of gravity) $\mathbf{y}$ of class y is defined by the *Mahalanobis* or *generalized distance*:

$$d(a, y) = (\mathbf{a} - \mathbf{y})'T^{-1}(\mathbf{a} - \mathbf{y}).$$

Vector $\mathbf{a}$ is assigned to the class y such that $d(a,y)$ is minimal over all groups. In the two-group case, we have that $\mathbf{a}$ is assigned to group y_1 if

$$d(a, y_1) < d(a, y_2).$$

Equality in the above may be resolved in accordance with user choice. Writing out explicitly the Euclidean distances associated with the matrix T^{-1}, and following some simplifications, we find that vector $\mathbf{a}$ is assigned to group y_1 if

$$(\mathbf{y}_1 - \mathbf{y}_2)'T^{-1}\mathbf{a} > \frac{1}{2}(\mathbf{y}_1 - \mathbf{y}_2)'T^{-1}(\mathbf{y}_1 + \mathbf{y}_2)$$

and to group y_2 if

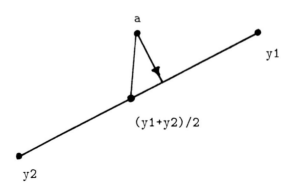

Figure 4.2: The assignment of a new sample **a** to one of two groups of centres $\mathbf{y}_1$ and $\mathbf{y}_2$.

$$(\mathbf{y}_1 - \mathbf{y}_2)'T^{-1}\mathbf{a} < \frac{1}{2}(\mathbf{y}_1 - \mathbf{y}_2)'T^{-1}(\mathbf{y}_1 + \mathbf{y}_2).$$

The left hand side is the T^{-1}-projection of **a** onto $\mathbf{y}_1 - \mathbf{y}_2$ (i.e. the vector connecting $\mathbf{y}_2$ to $\mathbf{y}_1$; and the right hand side is the T^{-1}-projection of $(\mathbf{y}_1 + \mathbf{y}_2)/2$ onto $\mathbf{y}_1 - \mathbf{y}_2$ (see Figure 4.2). This concords well with an intuitive idea of the objective in assignment: a new sample is assigned to a group if it lies closer to it than does the mid-way point between the two groups.

This allocation rule may be rewritten as

$$(\mathbf{y}_1 - \mathbf{y}_2)'T^{-1}(\mathbf{a} - \frac{\mathbf{y}_1 + \mathbf{y}_2}{2}) > 0 \Longrightarrow \mathbf{a} \to y_1$$

$$(\mathbf{y}_1 - \mathbf{y}_2)'T^{-1}(\mathbf{a} - \frac{\mathbf{y}_1 + \mathbf{y}_2}{2}) < 0 \Longrightarrow \mathbf{a} \to y_2.$$

The left hand side here is known as *Fisher's linear discriminant function*.

4.2.3 Bayesian Discrimination: Quadratic Case

The assignment aspect of discrimination is at issue in this section. As a general rule, it is clearly best if we attempt to take as much information as possible about the problem into account. Let us look at how successively more problem–related characteristics can be considered, but simultaneously we will have to pinpoint

4.2. MATHEMATICAL DESCRIPTION

difficulties in the implementation of our successive solutions. Overcoming these difficulties will lead to other approaches for the assignment problem, as will be seen.

Consider a vector of measured parameters, $\mathbf{x}$, relating to attributes of galaxies. Next consider that a sample of galaxies which is being studied consists of 75% spirals and 25% ellipticals. That is

$$P(spiral) = 0.75,$$
$$P(elliptical) = 0.25.$$

where $P(.)$ denotes probability. In the absence of any other information, we would therefore assign any unknown galaxy to the class of spirals. In the long run, we would be correct in 75% of cases, but we have obviously derived a very crude assignment rule.

Consider now that we are given also the conditional probabilities: for a particular set of parameter values, $\mathbf{x}_0$, we have

$$P(spiral \mid \mathbf{x}_0) = 0.3$$
$$P(elliptical \mid \mathbf{x}_0) = 0.7.$$

In this case, we are led to choose the class of ellipticals for our unknown galaxy, for which we have measured the parameter values $\mathbf{x}_0$. The conditional probabilities used above are referred to as *prior* or *a priori* probabilities.

This leads to Bayes' rule for the assignment of an unknown object to group c rather than to any other group, y:

$$P(c \mid \mathbf{x}_0) > P(y \mid \mathbf{x}_0) \quad for\ all\ y \neq c. \tag{4.1}$$

Tied optimal assignment values may be arbitrarily resolved.

A difficulty arises with Bayes' rule as defined above: although we could attempt to determine $P(c \mid \mathbf{x})$ for all possible values of $\mathbf{x}$ (or, perhaps, for a discrete set of such values), this is cumbersome. In fact, it is usually simpler to derive values for $P(\mathbf{x}_0 \mid c)$, i.e. the probability of having a given set of measurements, $\mathbf{x}_0$, given that we are dealing with a given class, c. Such probabilities are referred to as *posterior* or *a posteriori* probabilities. Bayes' theorem relates priors and posteriors. We have:

$$P(c \mid \mathbf{x}_0) = \frac{P(\mathbf{x}_0 \mid c) P(c)}{\sum_{all\ y} P(\mathbf{x}_0 \mid y) P(y)}. \tag{4.2}$$

All terms on the right hand side can be sampled: $P(c)$ is determined straightforwardly; $P(\mathbf{x}_0 \mid c)$ may be sampled by looking at each parameter in turn among the vector $\mathbf{x}_0$, and deriving estimates for the members of class c.

Substituting equation (4.2) into both sides of equation (4.1), and cancelling the common denominator, gives an assignment rule of the following form: *choose class c over all classes y, if*

$$P(\mathbf{x}_0 \mid c) \, P(c) > P(\mathbf{x}_0 \mid y) \, P(y) \quad for\ all\ y \neq c. \tag{4.3}$$

This form of Bayes' rule is better than the previous one. But again a difficulty arises: a great deal of sampling is required to estimate the terms of expression (4.3). Hence it is convenient to make distributional assumptions about the data. As always the Gaussian or normal distribution (for historical rectitude, known in the French literature as the distribution of Gauss–Laplace) figures prominently.

The multivariate normal density function (defining a multidimensional bell-shaped curve) is taken to better represent the distribution of $\mathbf{x}$ than the single point as heretofore. This is defined as

$$(2\pi)^{-\frac{n}{2}} \mid V \mid^{-\frac{1}{2}} exp\, (-\frac{1}{2}(\mathbf{x}-\mathbf{g})'V^{-1}(\mathbf{x}-\mathbf{g}))$$

where V is the variance–covariance matrix. It is of dimensions $m \times m$, if m is the dimensionality of the space. If equal to the identity matrix, it would indicate that $\mathbf{x}$ is distributed in a perfectly symmetric fashion with no privileged direction of elongation. $\mid V \mid$ is the determinant of the matrix V.

Assuming that each group, c, is a Gaussian, we have

$$P(\mathbf{x} \mid c) = (2\pi)^{-\frac{n}{2}} \mid V_c \mid^{-\frac{1}{2}} exp\, (-\frac{1}{2}(\mathbf{x}-\mathbf{g}_c)'V_c^{-1}(\mathbf{x}-\mathbf{g}_c))$$

where $\mathbf{g}_c$ is the centre of class c, and V_c is its variance–covariance matrix.

Substituting this into equation (4.3), taking natural logs of both sides of the inequality, and cancelling common terms on both sides, gives the following assignment rule: *assign $\mathbf{x}$ to class c if*

$$ln \mid V_c \mid + (\mathbf{x}-\mathbf{g}_c)'V_c^{-1}(\mathbf{x}-\mathbf{g}_c) - ln\, P(c)$$
$$< ln \mid V_y \mid + (\mathbf{x}-\mathbf{g}_y)'V_y^{-1}(\mathbf{x}-\mathbf{g}_y) - ln\, P(y) \quad for\ all\ y \neq c.$$

This expression is simplified by defining a "discriminant score" as

$$\delta_c(\mathbf{x}) = ln \mid V_c \mid + (\mathbf{x}-\mathbf{g_c})'V_c^{-1}(\mathbf{x}-\mathbf{g_c}).$$

The assignment rule then becomes: *assign $\mathbf{x}$ to class c if*

$$\delta_c(\mathbf{x}) - ln\, P(c) < \delta_y(\mathbf{x}) - ln\, P(y) \quad for\ all\ y \neq c.$$

The dividing curve between any two classes immediately follows from this. It is defined by

4.2. MATHEMATICAL DESCRIPTION

$$\delta_c(\mathbf{x}) - \ln P(c) = \delta_y(\mathbf{x}) - \ln P(y) \quad \text{for all } y \neq c$$

The shape of a curve defined by this equation is quadratic. Hence this general form of Bayesian discrimination is also referred to as *quadratic discrimination*.

4.2.4 Maximum Likelihood Discrimination

In a practical context we must estimate the mean vectors $(\mathbf{g}_y)$ and the variance–covariance matrices (V_y) from the data which may be taken to constitute a sample from an underlying population. These are termed *plug in estimators*, since they are sample estimates of the unknown parameters.

We have used a multivariate normal density function for $P(\mathbf{x} \mid y)$. If all n objects $\mathbf{x}_i$ have been independently sampled, then their joint distribution is

$$\mathcal{L} = \Pi_{i=1}^{n} P(\mathbf{x}_i \mid y).$$

Considering $\mathcal{L}$ as a function of the unknown parameters $\mathbf{g}$ and V, it is termed a *likelihood function*. The *principle of maximum likelihood* then states that we should choose the unknown parameters such that $\mathcal{L}$ is maximized. The classical approach for optimizing $\mathcal{L}$ is to differentiate it with respect to $\mathbf{g}$ and then with respect to V, and to set the results equal to zero. Doing this for the multivariate normal expression used previously allows us to derive estimates for the mean and covariances as follows.

$$\hat{\mathbf{g}} = \frac{1}{n} \sum_{i=1}^{n} \mathbf{x}_i$$

$$\hat{V} = \frac{1}{n} \sum_{i=1}^{n} (\mathbf{x}_i - \mathbf{g})(\mathbf{x}_i - \mathbf{g})'.$$

These are used to provide *maximum likelihood estimates* for the Bayesian classifier.

In a more general setting, we could wish to consider a multivariate normal mixture of the following form:

$$P(\mathbf{x} \mid y) = \sum_{k} w_k f_k(\mathbf{x} \mid \mathbf{g}_k, V_k)$$

where k ranges over the set of mixture members, w is a weighting factor, and the function f depends on the mean and the covariance structure of the mixture members. For such density functions, an iterative rather than an analytic approach is used, although boundedness and other convergence properties are problemsome (see Hand, 1981).

4.2.5 Bayesian Equal Covariances Case

The groups we study will not ordinarily have the same covariance structure. However it may be possible to assume that this is the case, and here we study what this implies in Bayesian discrimination.

The discriminant score, defined in section 4.2.3, when expanded is

$$\delta_c(\mathbf{x}) = ln \mid V_c \mid + \mathbf{x}' V_c^{-1} \mathbf{x} - \mathbf{g}_c' V_c^{-1} \mathbf{x} - \mathbf{x}' V_c^{-1} \mathbf{g}_c + \mathbf{g}_c' V_c^{-1} \mathbf{g}_c.$$

The first two terms on the right hand side can be ignored since they will feature on both sides of the assignment rule (by virtue of our assumption of equal covariances); and the third and fourth terms are equal. If we write

$$\phi_c(\mathbf{x}) = 2\mathbf{g}_c' V_c^{-1} \mathbf{x} - \mathbf{g}_c' V_c^{-1} \mathbf{g}_c$$

then the assignment rule is: *assign* $\mathbf{x}$ *to class c if*

$$\phi_c(\mathbf{x}) + ln\ P(c) > \phi_y(\mathbf{x}) + ln\ P(y) \quad for\ all\ y \neq c.$$

However, ϕ can be further simplified. Its second term is a constant for a given group, a_0; and its first term can be regarded as a vector of constant coefficients (for a given group), $\mathbf{a}$. Hence ϕ may be written as

$$\phi_c(\mathbf{x}) = a_{c0} + \sum_{j=1}^{m} a_{cj} x_j.$$

Assuming $P(c) = P(y)$, for all y, the assignment rule in the case of equal covariances thus involves a linear decision surface. We have a result which is particularly pleasing from the mathematical point of view: Bayesian discrimination in the equal covariances case, when the group cardinalities are equal, gives *exactly* the same decision rule (i.e. a linear decision surface) as linear discriminant analysis discussed from a geometric standpoint.

4.2.6 Non–Parametric Discrimination

Non–parametric (distribution–free) methods dispense with the need for assumptions regarding the probability density function. They have become very popular especially in the image processing area.

Given a vector of parameter values, $\mathbf{x}_0$, the probability that any unknown point will fall in a local neighbourhood of $\mathbf{x}_0$ may be defined in terms of the relative volume of this neighbourhood. If n' points fall in this region, out of a set of n points in total, and if v is the volume taken up by the region, then the probability that any unknown point falls in the local neighbourhood of $\mathbf{x}_0$ is n'/nv. An approach to classification arising out of this is as follows.

4.2. MATHEMATICAL DESCRIPTION

In the *k-NN* (*k* nearest neighbours) approach, we specify that the volume is to be defined by the k NNs of the unclassified point. Consider n_c of these k NNs to be members of class c, and n_y to be members of class y (with $n_c + n_y = k$). The conditional probabilities of membership in classes c and y are then

$$P(\mathbf{x}_0 \mid c) = \frac{n_c}{nv}$$

$$P(\mathbf{x}_0 \mid y) = \frac{n_y}{nv}.$$

Hence the decision rule is: *assign to group c if*

$$\frac{n_c}{nv} > \frac{n_y}{nv}$$

i.e. $n_c > n_y$.

The determining of NNs of course requires the definition of distance: the Euclidean distance is usually used.

An interesting theoretical property of the NN–rule relates it to the Bayesian misclassification rate. The latter is defined as

$$1 - max_y P(y \mid \mathbf{x}_0)$$

or, using notation introduced previously,

$$1 - P(c \mid \mathbf{x}_0). \qquad (4.4)$$

This is the probability that $\mathbf{x}_0$ will be misclassified, given that it should be classified into group c.

In the 1-NN approach, the misclassification rate is the product of: the conditional probability of class y given the measurement vector $\mathbf{x}$, and one minus the conditional probability of class y given the NN of $\mathbf{x}$ as the measurement vector:

$$\sum_{all\ y} P(y \mid \mathbf{x})(1 - P(y \mid NN(\mathbf{x}))). \qquad (4.5)$$

This is the probability that we assign to class y given that the NN is not in this class. It may be shown that the misclassification rate in the 1–NN approach (expression 4.5) is not larger than twice the Bayesian misclassification rate (expression 4.4). Hand (1981) may be referred to for the proof.

4.3 Examples and Bibliography

4.3.1 Practical Remarks

We can evaluate *error rates* by means of a training sample (to construct the discrimination surface) and a test sample. An optimistic error rate is obtained by reclassifying the design set: this is known as the *apparent error rate*. If an independent test sample is used for classifying, we arrive at the *true error rate*.

The *leaving one out* method attempts to use as much of the data as possible: for every subset of $n - 1$ objects from the given n objects, a classifier is designed, and the object omitted is assigned. This leads to the overhead of n discriminant analyses, and to n tests from which an error rate can be derived. Another approach to appraising the results of a discriminant analysis is to determine a *confusion matrix* which is a contingency table (a table of frequencies of co–occurrence) crossing the known groups with the obtained groups.

We may improve our discrimination by implementing a *reject option*: if for instance we find $P(\mathbf{x} \mid c) > P(\mathbf{x} \mid y)$ for all groups $y \neq c$, we may additionally require that $P(\mathbf{x} \mid c)$ be greater than some threshold for assignment of $\mathbf{x}$ to c. Such an approach will of course help to improve the error rate.

There is no best discrimination method. A few remarks concerning the advantages and disadvantages of the methods studied are as follows.

- Analytical simplicity or computational reasons may lead to initial consideration of linear discriminant analysis or the NN–rule.

- Linear discrimination is the most widely used in practice. Often the 2-group method is used repeatedly for the analysis of pairs of multigroup data (yielding $k(k-1)/2$ decision surfaces for k groups).

- To estimate the parameters required in quadratic discrimination requires more computation and data than in the case of linear discrimination. If there is not a great difference in the group covariance matrices, then the latter will perform as well as quadratic discrimination.

- The k–NN rule is simply defined and implemented, especially if there is insufficient data to adequately define sample means and covariance matrices.

- MDA is most appropriately used for *feature selection*. As in the case of PCA, we may want to focus on the variables used in order to investigate the differences between groups; to create synthetic variables which improve the grouping ability of the data; to arrive at a similar objective by discarding irrelevant variables; or to determine the most parsimonious variables for graphical representational purposes.

4.3.2 Examples from Astronomy

1. H.-M. Adorf, "Classification of low–resolution stellar spectra via template matching — a simulation study", *Data Analysis and Astronomy*, (Proceedings of International Workshop on Data Analysis and Astronomy, Erice, Italy, April 1986) Plenum Press, New York, 1986 (in press).

2. E. Antonello and G. Raffaelli, "An application of discriminant analysis to variable and nonvariable stars", *Publications of the Astronomical Society of the Pacific*, **95**, 82–85, 1983.

    (Multiple Discriminant Analysis is used.)

3. M. Fracassini, L.E. Pasinetti and G. Raffaelli, "Discriminant analysis on pulsar groups in the diagram $\dot{P}$ versus P", in *Proceedings of a Course and Workshop on Plasma Astrophysics*, European Space Agency Special Publication 207, 315–317, 1984.

4. M. Fracassini, P. Maggi, L.E. Pasinetti and G. Raffaelli, "Pair of pulsars in the diagram $\dot{P}$ versus P", *Proceedings of the Joint Varenna–Abastumani International School and Workshop on Plasma Astrophysics*, Sukhami, European Space Agency Special Publication 251, 441–445, 1986.

5. A. Heck, "An application of multivariate statistical analysis to a photometric catalogue", *Astronomy and Astrophysics*, **47**, 129–135, 1976.

    (Multiple Discriminant Analysis and a stepwise procedure are applied.)

6. J.F. Jarvis and J.A. Tyson, "FOCAS — Faint object classification and analysis system", *SPIE Instrumentation in Astronomy III*, **172**, 422–428, 1979.

    (See also other references of Tyson/Jarvis and Jarvis/Tyson.)

7. J.F. Jarvis and J.A. Tyson, "Performance verification of an automated image cataloging system", *SPIE Applications of Digital Image Processing to Astronomy*, **264**, 222–229, 1980.

8. J.F. Jarvis and J.A. Tyson, "FOCAS — Faint object classification and analysis system", *The Astronomical Journal*, **86**, 1981, 476–495.

    (A hyperplane separation surface is determined in a space defined by 6 parameters used to characterise the objects. This is a 2–stage procedure where the first stage is that of training, and the second stage uses a partitioning clustering method.)

9. M.J. Kurtz, "Progress in automation techniques for MK classification", in ed. R.F. Garrison, *The MK Process and Stellar Classification*, David Dunlap Observatory, University of Toronto, 1984, pp. 136–152.

(Essentially a k–NN approach is used for assigning spectra to known stellar spectral classes.)

10. H.T. MacGillivray, R. Martin, N.M. Pratt, V.C. Reddish, H. Seddon, L.W.G. Alexander, G.S. Walker, P.R. Williams, "A method for the automatic separation of the images of galaxies and stars from measurements made with the COSMOS machine", *Monthly Notices of the Royal Astronomical Society*, **176**, 265–274, 1976.

    (Different parameters are appraised for star/galaxy separation. Kurtz — see reference in Chapter 3 (Cluster Analysis) — lists other parameters which have been used for the same objective.)

11. M.L. Malagnini, "A classification algorithm for star–galaxy counts", in *Statistical Methods in Astronomy*, European Space Agency Special Publication 201, 1983, pp. 69–72.

    (A linear classifier is used and is further employed in the following reference.)

12. M.L. Malagnini, F. Pasian, M. Pucillo and P. Santin, "FODS: a system for faint object detection and classification in astronomy", *Astronomy and Astrophysics*, **144**, 1985, 49–56.

13. "Recommendations for Guide Star Selection System", GSSS Group documentation, Space Telescope Science Institute, Baltimore, 1984.

    (A Bayesian approach, using the IMSL subroutine library — see below — is employed in the GSSS system.)

14. W.J. Sebok, "Optimal classification of images into stars or galaxies — a Bayesian approach", *The Astronomical Journal*, **84**, 1979, 1526–1536.

    (The design of a classifier, using galaxy models, is studied in depth and validated on Schmidt plate data.)

15. J.A. Tyson and J.F. Jarvis, "Evolution of galaxies: automated faint object counts to 24th magnitude", *The Astrophyiscal Journal*, **230**, 1979, L153–L156.

    (A continuation of the work of Jarvis and Tyson, 1979, above.)

16. F. Valdes, "Resolution classifier", *SPIE Instrumentation in Astronomy IV*, **331**, 1982, 465–471.

    (A Bayesian classifier is used, which differs from that used by Sebok, referenced above. The choice is thoroughly justified. A comparison is also made with the hyperplane fitting method used in the FOCAS system — see the references of Jarvis and Tyson. It is concluded that the results obtained

4.3. EXAMPLES AND BIBLIOGRAPHY

within the model chosen are better than a hyperplane based approach in parameter space; but that the latter is computationally more efficient.)

4.3.3 General References

In the following, some software packages are included. The accompanying documentation often constitutes a quick and convenient way to get information on analysis methods.

1. S.-T. Bow, *Pattern Recognition*, Marcel Dekker, New York, 1984.

   (A textbook detailing a range of Discriminant Analysis methods, together with clustering and other topics.)

2. C. Chatfield and A.J. Collins, *Introduction to Multivariate Analysis*, Chapman and Hall, London, 1980.

   (A good introductory textbook.)

3. E. Diday, J. Lemaire, J. Pouget and F. Testu, *Eléments d'Analyse de Données*, Dunod, Paris, 1982.

   (Describes a large range of methods.)

4. R. Duda and P. Hart, *Pattern Classification and Scene Analysis*, Wiley, New York, 1973.

   (Excellent treatment of many image processing problems.)

5. R.A. Fisher, "The use of multiple measurements in taxonomic problems", *The Annals of Eugenics*, **7**, 179–188, 1936.

   (Still an often referenced paper; contains the famous Iris data.)

6. K. Fukunaga, *Introduction to Statistical Pattern Recognition*, Academic Press, New York, 1972.

7. D.J. Hand, *Discrimination and Classification*, Wiley, New York, 1981.

   (A comprehensive description of a wide range of methods; very recommendable.)

8. International Mathematical and Statistical Library (IMSL), Manual sections on ODFISH, ODNORM, etc.

   (A useful range of algorithms is available in this widely used subroutine library.)

9. M. James, *Classification Algorithms*, Collins, London, 1985.

   (A readable introduction.)

10. M.G. Kendall, *Multivariate Analysis*, Griffin, London, 1980 (2nd ed.).

    (Dated in relation to computing techniques, but exceptionally clear and concise in its treatment of many practical problems.)

11. P.A. Lachenbruch, *Discriminant Analysis*, Hafner Press, New York, 1975.

12. J.L. Melsa and D.L. Cohn, *Decision and Estimation Theory*, McGraw–Hill, New York, 1978.

    (A readable decision theoretic perspective.)

13. J.M. Romeder, *Méthodes et Programmes d'Analyse Discriminante*, Dunod, Paris, 1973.

    (A survey of commonly-used techniques.)

14. Statistical Analysis System (SAS), SAS Institute Inc., Box 8000, Cary, NC 27511–8000, USA; Manual chapters on STEPDISC, NEIGHBOUR, etc.

    (A range of relevant algorithms is available in this, — one of the premier statistical packages.)

4.4 Software and Sample Implementation

The first section which follows concerns Linear Discriminant Analysis. The subroutine LDA calls:

1. OUTMAT for matrix printing.

2. MATINV for matrix inversion.

3. OUTPRX for outputting projections on the discriminant axis.

4. PROJX for determining these projections.

The next section which follows concerns Multiple Discriminant Analysis. The subroutine MDA calls:

1. MATINV for matrix inversion.

2. TRED2 for reducing a matrix to tridiagonal form.

3. TQL2 for determining the eigenvalues and eigenvectors of a tridiagonal matrix.

4. OUTEVL for outputting eigenvalues.

5. OUTEVC for outputting eigenvectors.

6. OUTPRX for outputting projections of row–points.

7. OUTPRY for outputting projections of column–points.

8. PROJX for determining projections of row–points.

9. PROJY for determining projections of column–points.

In order to avoid difficulties with the matrix inversion routine, it may be advisable to multiply all input data values by a constant — a non-standard, but reasonable, way to avoid singularity in the case of very small values.

The third section which follows concerns K–NNs Discriminant Analysis. The principal subroutine here is KNN.

All three methods require driver programs, only, to run.

4.4.1 Program Listing: Linear Discriminant Analysis

```
C++++++++++++++++++++++++++++++++++++++++++++++++++++++++
C
C   Carry out a LINEAR DISCRIMINANT ANALYSIS, assigning ungrouped
C   items to the closest group centre, using the Mahalanobis
C   distance.
C
C
C   To call:    CALL LDA(N,M,DATA,GP,IPRINT,MEAN,MGP,TOTAL,DIFFVC,
C                         W1,W2,IERR)      where
C
C
C   N, M   : integer dimensions of ...
C   DATA   : input data (real).
C            On output the first column of DATA contains projections
C            of all N items on the line connecting the two group
C            means. Zero is the boundary point.
C   GP     : Integer vector of length N giving group assignments. An
C            unassigned item has group 0. Otherwise groups 1 and 2
C            will be looked for, and other values here are not
C            acceptable.
C   IPRINT : integer; print options (= 3: full; otherwise none).
C   MEAN   : real vector of length M (number of attributes or
C            variables).
C   MGP    : real array of dimensions 2 by M.
C   TOTAL  : real array of dims. M by M; on output contains inverse
C            of total variance/covariance matrix.
C   DIFFVC : real vector of length M.
C   W1, W2 : real vectors of length M.
C
C
C   Inputs here are N, M, DATA, GP, IPRINT (and IERR).
C   The principle output information is contained in DATA.
C   IERR = 1 means that more than two groups have been specified;
C   IERR = 2 means that the total variance-covariance matrx was
C            singular.
C
C   Note: we require N > M > 2, to prevent the seeking of the
C         inverse of a singular matrix.
C
C----------------------------------------------------------------
        SUBROUTINE LDA(N,M,DATA,GP,IPRINT,MEAN,MGP,TOTAL,DIFFVC,
     X                                          W1,W2,IERR)
        REAL    DATA(N,M), TOTAL(M,M), MEAN(M), MGP(2,M)
        REAL    DIFFVC(M), W1(M), W2(M)
        INTEGER GP(N), NOG(2)
C
C       Form global mean.
C
```

4.4. SOFTWARE AND SAMPLE IMPLEMENTATION

```
          IERR = 0
          NEFF = 0
          DO 50 I = 1, N
             IF (GP(I).NE.0) NEFF = NEFF + 1
             IF (GP(I).LE.2) GOTO 40
                IERR = 1
                GOTO 9000
   40     CONTINUE
   50     CONTINUE
C
          DO 200 J = 1, M
             MEAN(J) = 0.0
             DO 100 I = 1, N
                IF (GP(I).NE.0) MEAN(J) = MEAN(J) + DATA(I,J)
  100        CONTINUE
             MEAN(J) = MEAN(J)/FLOAT(NEFF)
  200     CONTINUE
C
C         Form (total) variance-covariance matrix.
C
          DO 500 J1 = 1, M
             DO 400 J2 = 1, M
                TOTAL(J1,J2) = 0.0
                DO 300 I = 1, N
                   IF (GP(I).NE.0) TOTAL(J1,J2) = TOTAL(J1,J2) +
     X               (DATA(I,J1)-MEAN(J1))*(DATA(I,J2)-MEAN(J2))
  300           CONTINUE
                TOTAL(J1,J2) = TOTAL(J1,J2)/FLOAT(NEFF)
  400        CONTINUE
  500     CONTINUE
C
          IMAT = 1
          IF (IPRINT.EQ.3) CALL OUTMAT(IMAT,M,TOTAL)
C
C         Form group means.
C
          DO 700 J = 1, M
             DO 600 K = 1, 2
                MGP(K,J) = 0.0
  600        CONTINUE
  700     CONTINUE
C
          DO 900 I = 1, N
             G = GP(I)
             IF (G.EQ.0) GOTO 900
             NOG(G) = NOG(G) + 1
             DO 800 J = 1, M
                MGP(G,J) = MGP(G,J) + DATA(I,J)
  800        CONTINUE
```

```
       900    CONTINUE
C
              DO 1100 K = 1, 2
                 DO 1000 J = 1, M
                    MGP(K,J) = MGP(K,J)/NOG(K)
      1000        CONTINUE
      1100     CONTINUE
C
C             Invert variance-covariance matrix.
C
              CALL MATINV(M,TOTAL,D,W1,W2)
              IF (D.GT.0.00001) GOTO 1150
                 IERR = 2
                 GOTO 9000
      1150    CONTINUE
C
C             Form difference vector of group mean vectors.
C
              DO 1200 J = 1, M
                 DIFFVC(J) = MGP(1,J) - MGP(2,J)
                 MEAN(J) = (MGP(1,J) + MGP(2,J))/2.
      1200    CONTINUE
C
C             Determine projections and output them.
C
              CALL PROJX(N,M,DATA,MEAN,W1,TOTAL,DIFFVC)
              IF (IPRINT.EQ.3) CALL OUTPRX(N,M,DATA)
C
C
C
      9000    CONTINUE
              RETURN
              END
C++++++++++++++++++++++++++++++++++++++++++++++++++++++++++
C
C  Output a matrix.
C
C----------------------------------------------------------
              SUBROUTINE OUTMAT(IMAT,M,ARRAY)
C
C                Output array.
C
              DIMENSION ARRAY(M,M)
C
              IF (IMAT.EQ.1)WRITE (6,900)
              DO 100 K1 = 1, M
                 WRITE (6,1000) (ARRAY(K1,K2),K2=1,M)
       100    CONTINUE
C
```

4.4. SOFTWARE AND SAMPLE IMPLEMENTATION

```
   900   FORMAT(' VARIANCE/COVARIANCE MATRIX FOLLOWS.',/)
  1000   FORMAT(10(2X,F8.4))
         RETURN
         END
C++++++++++++++++++++++++++++++++++++++++++++++++++++++++
C
C  Invert a symmetric matrix and calculate its determinant.
C
C
C  To call:     CALL MATINV(M,ARRAY,DET,W1,W2)    where
C
C
C  M       : dimension of ...
C  ARRAY   : input matrix which is replaced by its inverse.
C  NORDER  : degree of matrix (order of determinant)
C  DET     : determinant of input matrix.
C  W1, W2  : work vectors of dimension M.
C
C
C  Reference: Philip B Bevington, "Data Reduction and Error
C             Analysis for the Physical Sciences", McGraw-Hill,
C             New York, 1969, pp. 300-303.
C
C---------------------------------------------------------------
         SUBROUTINE MATINV(M,ARRAY,DET,IK,JK)
         REAL    ARRAY(M,M), IK(M), JK(M)
C
    10   DET = 1.0
    11   DO 100 K = 1, M
C        Find largest element ARRAY(I,J) in rest of matrix.
         AMAX = 0.0
    21      DO 30 I = K, M
               DO 30 J = K, M
    23            IF (ABS(AMAX)-ABS(ARRAY(I,J))) 24,24,30
    24            AMAX = ARRAY(I,J)
                  IK(K) = I
                  JK(K) = J
    30      CONTINUE
C        Interchange rows and columns to put AMAX in ARRAY(K,K).
    31      IF (AMAX) 41,32,41
    32      DET = 0.0
            GOTO 140
    41      I = IK(K)
            IF (I-K) 21,51,43
    43      DO 50 J = 1, M
               SAVE = ARRAY(K,J)
               ARRAY(K,J) = ARRAY(I,J)
    50      ARRAY(I,J) = -SAVE
    51      J = JK(K)
            IF (J-K) 21,61,53
    53      DO 60 I = 1, M
```

```
              SAVE = ARRAY(I,K)
              ARRAY(I,K) = ARRAY(I,J)
      60      ARRAY(I,J) = -SAVE
C             Accumulate elements of inverse matrix.
      61      DO 70 I = 1, M
              IF (I-K) 63,70,63
      63      ARRAY(I,K) = -ARRAY(I,K)/AMAX
      70      CONTINUE
      71      DO 80 I = 1, M
              DO 80 J = 1, M
              IF (I-K) 74,80,74
      74      IF (J-K) 75,80,75
      75      ARRAY(I,J) = ARRAY(I,J) + ARRAY(I,K)*ARRAY(K,J)
      80      CONTINUE
      81      DO 90 J = 1, M
              IF (J-K) 83,90,83
      83      ARRAY(K,J) = ARRAY(K,J)/AMAX
      90      CONTINUE
              ARRAY(K,K) = 1.0/AMAX
      100     DET = DET * AMAX
C             Restore ordering of matrix.
      101     DO 130 L = 1, M
              K = M - L + 1
              J = IK(K)
              IF (J-K) 111,111,105
      105     DO 110 I = 1, M
              SAVE = ARRAY(I,K)
              ARRAY(I,K) = -ARRAY(I,J)
      110     ARRAY(I,J) = SAVE
      111     I = JK(K)
              IF (I-K) 130,130,113
      113     DO 120 J = 1, M
              SAVE = ARRAY(K,J)
              ARRAY(K,J) = -ARRAY(I,J)
      120     ARRAY(I,J) = SAVE
      130     CONTINUE
      140     RETURN
              END
C++++++++++++++++++++++++++++++++++++++++++++++++++++++++++++
C
C  Output projections of row points.
C
C-------------------------------------------------------------
              SUBROUTINE OUTPRX(N,M,PRJN)
              REAL    PRJN(N,M)
C
C
```

4.4. SOFTWARE AND SAMPLE IMPLEMENTATION

```
          NUM = 1
          WRITE (6,1000)
          WRITE (6,1010)
          WRITE (6,1020)
          DO 100 K = 1, N
             WRITE (6,1030) K,(PRJN(K,J),J=1,NUM)
   100    CONTINUE
C
  1000    FORMAT(1H0,'PROJECTIONS OF ROW-POINTS FOLLOW.',/)
  1010    FORMAT(' OBJECT    PROJN')
  1020    FORMAT(' ------    ------')
  1030    FORMAT(I5,2X,F8.4)
          RETURN
          END
C++++++++++++++++++++++++++++++++++++++++++++++++++++++++++
C
C  Form projections of row-points on factors.
C
C------------------------------------------------------------
          SUBROUTINE PROJX(N,M,DATA,MEAN,VEC,TOTINV,DIFF)
          REAL    DATA(N,M), MEAN(M), VEC(M), TOTINV(M,M), DIFF(M)
C
          NUM = 1
          DO 300 K = 1, N
             DO 50 L = 1, M
                VEC(L) = DATA(K,L)
    50       CONTINUE
             DO 200 I = 1, NUM
                DATA(K,I) = 0.0
                DO 100 J1 = 1, M
                   DO 75 J2 = 1, M
                      DATA(K,I) = DATA(K,I) + (VEC(J1)-MEAN(J1))*
        X                         TOTINV(J1,J2)*DIFF(J2)
    75             CONTINUE
   100          CONTINUE
   200       CONTINUE
   300    CONTINUE
C
          RETURN
          END
```

4.4.2 Program Listing: Multiple Discriminant Analysis

```
C+++++++++++++++++++++++++++++++++++++++++++++++++++++++++++
C
C  Carry out a MULTIPLE DISCRIMINANT ANALYSIS
C              (DISCRIMINANT FACTOR ANALYSIS,
C               CANONICAL DISCRININANT ANALYSIS).
C
C
C  To call:    CALL MDA(N,M,NG,DATA,GP,IPRINT,NOG,MEAN,MGP,TOTAL,
C                       BETWEE,BETW2,CPROJ,W1,W2,IERR)     where
C
C
C  N, M   : integer dimensions of ...
C  DATA   : input data (real).
C            On output the first NG-1 columns of DATA contain projns.
C            of the N items on the discriminant factors.
C  NG     : (integer) number of groups.
C  GP     : Integer vector of length N giving group assignments.
C            Must be specified correctly - no 0s or values > NG.
C  IPRINT : integer; print options (= 3: full; otherwise none).
C  NOG    : integer vect. of len. NG (to contain gp. cardinalities).
C  MEAN   : real vector of length M (number of attributes or vbes).
C  MGP    : real array of dimensions 2 by M.
C  TOTAL  : real array of dims. M by M; on output contains inverse
C            of total variance/covariance matrix.
C  BETWEE : real array of dimensions NG by NG.
C  BETW2  : real array of dimensions NG by NG.
C  CPROJ  : real array of dimensions M by NG; on output contains the
C            coefficients of the discriminant factors in terms of the
C            original variables.
C  W1, W2 : real vectors of length M.
C  IERR   : initially 0; = 1 if there is no convergence in the TQL2
C            eigenroutine; = 2 if the total variance-covariance to be
C            inverted is singular, - in this case, check that there
C            are no columns with identical values, that N > M > NG,
C            etc.
C
C  Inputs here are N, M, NG, DATA, GP, IPRINT (and IERR).
C  The principle output information is contained in DATA and CPROJ;
C  and W1(NG-1), W1(NG-2) ... contain the eigenvalues in decreasing
C  order.
C
C  Notes: we require that N > M > NG (otherwise, an error is likely
C          in the matrix inversion routine due to singularity).
C          NG-1 eigenvalues eigenvectors are output.
C
C-----------------------------------------------------------------
       SUBROUTINE MDA(N,M,NG,DATA,GP,IPRINT,NOG,MEAN,MGP,TOTAL,
      X               BETWEE,BETW2,CPROJ,W1,W2,IERR)
```

4.4. SOFTWARE AND SAMPLE IMPLEMENTATION

```
        REAL    DATA(N,M), TOTAL(M,M), MEAN(M), MGP(NG,M)
        REAL    W1(M), W2(M), BETW2(NG,NG), CPROJ(M,NG)
        REAL    BETWEE(NG,NG)
        INTEGER GP(N), NOG(NG)
C
C       Form global mean.
C
        DO 200 J = 1, M
           MEAN(J) = 0.0
           DO 100 I = 1, N
              MEAN(J) = MEAN(J) + DATA(I,J)
  100      CONTINUE
           MEAN(J) = MEAN(J)/FLOAT(N)
  200   CONTINUE
C
C       Form (total) variance-covariance matrix.
C
        DO 500 J1 = 1, M
           DO 400 J2 = 1, M
              TOTAL(J1,J2) = 0.0
              DO 300 I = 1, N
                 TOTAL(J1,J2) = TOTAL(J1,J2) +
     X            (DATA(I,J1)-MEAN(J1))*(DATA(I,J2)-MEAN(J2))
  300         CONTINUE
              TOTAL(J1,J2) = TOTAL(J1,J2)/FLOAT(N)
  400      CONTINUE
  500   CONTINUE
C
        IMAT = 1
C       CALL OUTMAT(IMAT,M,TOTAL)
C
C       Form group means.
C
        DO 700 J = 1, M
           DO 600 K = 1, NG
              MGP(K,J) = 0.0
  600      CONTINUE
  700   CONTINUE
C
        DO 900 I = 1, N
           G = GP(I)
           IF (G.EQ.0) GOTO 9000
           NOG(G) = NOG(G) + 1
           DO 800 J = 1, M
              MGP(G,J) = MGP(G,J) + DATA(I,J)
  800      CONTINUE
  900   CONTINUE
C
        DO 1100 K = 1, NG
```

```
              DO 1000 J = 1, M
                 MGP(K,J) = MGP(K,J)/NOG(K)
 1000         CONTINUE
 1100      CONTINUE
C
C          Invert variance-covariance matrix.
C
           CALL MATINV(M,TOTAL,D,W1,W2)
           IF (D.GT.0.000001) GOTO 1150
              IERR = 2
              GOTO 9000
 1150      CONTINUE
           IMAT = 2
C          CALL OUTMAT(IMAT,M,TOTAL)
C
C          Form the symmetric variant of the BETWEE-groups
C          variance-covariance matrix for diagonalization.
C
           DO 1200 K1 = 1, NG
            DO 1200 K2 = 1, NG
             BETWEE(K1,K2) = 0.0
              DO 1200 J1 = 1, M
               DO 1200 J2 = 1, M
                D1 = MGP(K1,J1) - MEAN(J1)
                D2 = FLOAT(NOG(K1))/FLOAT(N)
                D2 = SQRT(D2)
                D3 = MGP(K2,J2) - MEAN(J2)
                D4 = FLOAT(NOG(K2))/FLOAT(N)
                D4 = SQRT(D4)
                BETWEE(K1,K2) = BETWEE(K1,K2) +
     X                          (D1*D2)*TOTAL(J1,J2)*(D3*D4)
 1200      CONTINUE
C
           IMAT = 4
C          CALL OUTMAT(IMAT,M,TOTAL)
C
C          Carry out eigenreduction.
C
           NG2 = NG
           CALL TRED2(NG,NG2,BETWEE,W1,W2,BETW2)
           CALL TQL2(NG,NG2,W1,W2,BETW2,IERR)
           IF (IERR.NE.0) GOTO 9000
C
C          Output eigenvalues and eigenvectors.
C
           IF (IPRINT.GT.1) CALL OUTEVL(N,M,NG,W1)
           IF (IPRINT.GT.1) CALL OUTEVC(N,M,NG,BETW2,NG-1)
C
C          Convert eigenvectors in NG-space to those in M-space.
C
```

4.4. SOFTWARE AND SAMPLE IMPLEMENTATION

```
              DO 1300 J = 1, M
                 DO 1300 K = 1, NG
                    CPROJ(J,K) = 0.0
                    DO 1300 J2 = 1, M
                       DO 1300 K2 = 1, NG
                          D1 = MGP(K2,J2) - MEAN(J2)
                          D2 = FLOAT(NOG(K2))/FLOAT(N)
                          D1 = D1*SQRT(D2)
                          CPROJ(J,K)=CPROJ(J,K)+
     X                    TOTAL(J,J2)*D1*BETW2(K2,NG-K+1)
 1300    CONTINUE
         IF (IPRINT.GT.1) CALL OUTEVC(N,NG,M,CPROJ,NG-1)
C
C        Determine projections and output them.
C
         CALL PROJX(N,M,NG,DATA,MEAN,CPROJ,W2,TOTAL)
         IF (IPRINT.EQ.3) CALL OUTPRX(N,M,NG,DATA)
C
C
C
 9000    CONTINUE
         RETURN
         END
C-------------------------------------------------------------
         SUBROUTINE OUTMAT(IMAT,M,ARRAY)
         DIMENSION ARRAY(M,M)
C
         WRITE (6,900) IMAT
         DO 100 K1 = 1, M
            WRITE (6,1000) (ARRAY(K1,K2),K2=1,M)
  100    CONTINUE
C
  900    FORMAT(' IMAT =',I6)
 1000    FORMAT(10(2X,F8.4))
         RETURN
         END
C++++++++++++++++++++++++++++++++++++++++++++++++++++++++++++
C
C  Invert a symmetric matrix and calculate its determinant.
C
C
C  To call:     CALL MATINV(M,ARRAY,DET,W1,W2)    where
C
C
C  M       : dimension of ...
C  ARRAY   : input matrix which is replaced by its inverse.
C  NORDER  : degree of matrix (order of determinant)
C  DET     : determinant of input matrix.
```

```
C   W1, W2  : work vectors of dimension M.
C
C
C   Reference: Philip B Bevington, "Data Reduction and Error
C              Analysis for the Physical Sciences", McGraw-Hill,
C              New York, 1969, pp. 300-303.
C
C------------------------------------------------------------------
        SUBROUTINE MATINV(M,ARRAY,DET,IK,JK)
        REAL    ARRAY(M,M), IK(M), JK(M)
C
   10   DET = 1.0
   11   DO 100 K = 1, M
C       Find largest element ARRAY(I,J) in rest of matrix.
        AMAX = 0.0
   21       DO 30 I = K, M
               DO 30 J = K, M
   23             IF (ABS(AMAX)-ABS(ARRAY(I,J))) 24,24,30
   24             AMAX = ARRAY(I,J)
                  IK(K) = I
                  JK(K) = J
   30   CONTINUE
C       Interchange rows and columns to put AMAX in ARRAY(K,K).
   31   IF (AMAX) 41,32,41
   32   DET = 0.0
        GOTO 140
   41   I = IK(K)
        IF (I-K) 21,51,43
   43   DO 50 J = 1, M
            SAVE = ARRAY(K,J)
            ARRAY(K,J) = ARRAY(I,J)
   50   ARRAY(I,J) = -SAVE
   51   J = JK(K)
        IF (J-K) 21,61,53
   53   DO 60 I = 1, M
            SAVE = ARRAY(I,K)
            ARRAY(I,K) = ARRAY(I,J)
   60   ARRAY(I,J) = -SAVE
C       Accumulate elements of inverse matrix.
   61   DO 70 I = 1, M
            IF (I-K) 63,70,63
   63       ARRAY(I,K) = -ARRAY(I,K)/AMAX
   70   CONTINUE
   71   DO 80 I = 1, M
            DO 80 J = 1, M
               IF (I-K) 74,80,74
   74          IF (J-K) 75,80,75
   75          ARRAY(I,J) = ARRAY(I,J) + ARRAY(I,K)*ARRAY(K,J)
   80   CONTINUE
```

```
      81        DO 90 J = 1, M
                  IF (J-K) 83,90,83
      83          ARRAY(K,J) = ARRAY(K,J)/AMAX
      90        CONTINUE
                ARRAY(K,K) = 1.0/AMAX
     100        DET = DET * AMAX
C             Restore ordering of matrix.
     101        DO 130 L = 1, M
                  K = M - L + 1
                  J = IK(K)
                  IF (J-K) 111,111,105
     105          DO 110 I = 1, M
                    SAVE = ARRAY(I,K)
                    ARRAY(I,K) = -ARRAY(I,J)
     110          ARRAY(I,J) = SAVE
     111        I = JK(K)
                IF (I-K) 130,130,113
     113        DO 120 J = 1, M
                    SAVE = ARRAY(K,J)
                    ARRAY(K,J) = -ARRAY(I,J)
     120        ARRAY(I,J) = SAVE
     130      CONTINUE
     140      RETURN
             END
C+++++++++++++++++++++++++++++++++++++++++++++++++++++++++++++
C
C Reduce a real, symmetric matrix to a symmetric, tridiagonal
C matrix.
C
C To call:    CALL TRED2(NM,N,A,D,E,Z)    where
C
C NM = row dimension of A and Z;
C N = order of matrix A (will always be <= NM);
C A = symmetric matrix of order N to be reduced to tridiagonal
C     form;
C D = vector of dim. N containing, on output, diagonal elements of
C     tridiagonal matrix;
C E = working vector of dim. at least N-1 to contain subdiagonal
C     elements;
C Z = matrix of dims. NM by N containing, on output, orthogonal
C     transformation matrix producting the reduction.
C
C Normally a call to TQL2 will follow the call to TRED2 in order
C to produce all eigenvectors and eigenvalues of matrix A.
C
C Algorithm used: Martin et al., Num. Math. 11, 181-195, 1968.
C
C Reference: Smith et al., Matrix Eigensystem Routines - EISPACK
C Guide, Lecture Notes in Computer Science 6, Springer-Verlag,
```

```
C 1976, pp. 489-494.
C
C-----------------------------------------------------------------
      SUBROUTINE TRED2(NM,N,A,D,E,Z)
C
      REAL A(NM,N),D(N),E(N),Z(NM,N)
C
      DO 100 I = 1, N
         DO 100 J = 1, I
            Z(I,J) = A(I,J)
  100 CONTINUE
      IF (N.EQ.1) GOTO 320
      DO 300 II = 2, N
         I = N + 2 - II
         L = I - 1
         H = 0.0
         SCALE = 0.0
         IF (L.LT.2) GOTO 130
         DO 120 K = 1, L
            SCALE = SCALE + ABS(Z(I,K))
  120    CONTINUE
         IF (SCALE.NE.0.0) GOTO 140
  130    E(I) = Z(I,L)
         GOTO 290
  140    DO 150 K = 1, L
            Z(I,K) = Z(I,K)/SCALE
            H = H + Z(I,K)*Z(I,K)
  150    CONTINUE
C
         F = Z(I,L)
         G = -SIGN(SQRT(H),F)
         E(I) = SCALE * G
         H = H - F * G
         Z(I,L) = F - G
         F = 0.0
C
         DO 240 J = 1, L
            Z(J,I) = Z(I,J)/H
            G = 0.0
C           Form element of A*U.
            DO 180 K = 1, J
               G = G + Z(J,K)*Z(I,K)
  180       CONTINUE
            JP1 = J + 1
            IF (L.LT.JP1) GOTO 220
            DO 200 K = JP1, L
               G = G + Z(K,J)*Z(I,K)
  200       CONTINUE
C           Form element of P where P = I - U U' / H .
```

4.4. SOFTWARE AND SAMPLE IMPLEMENTATION

```
  220         E(J) = G/H
              F = F + E(J) * Z(I,J)
  240       CONTINUE
            HH = F/(H + H)
C           Form reduced A.
            DO 260 J = 1, L
              F = Z(I,J)
              G = E(J) - HH * F
              E(J) = G
              DO 250 K = 1, J
                Z(J,K) = Z(J,K) - F*E(K) - G*Z(I,K)
  250         CONTINUE
  260       CONTINUE
  290       D(I) = H
  300     CONTINUE
  320     D(1) = 0.0
          E(1) = 0.0
C         Accumulation of transformation matrices.
          DO 500 I = 1, N
            L = I - 1
            IF (D(I).EQ.0.0) GOTO 380
            DO 360 J = 1, L
              G = 0.0
              DO 340 K = 1, L
                G = G + Z(I,K) * Z(K,J)
  340         CONTINUE
              DO 350 K = 1, L
                Z(K,J) = Z(K,J) - G * Z(K,I)
  350         CONTINUE
  360       CONTINUE
  380       D(I) = Z(I,I)
            Z(I,I) = 1.0
            IF (L.LT.1) GOTO 500
            DO 400 J = 1, L
              Z(I,J) = 0.0
              Z(J,I) = 0.0
  400       CONTINUE
  500     CONTINUE
C
          RETURN
          END
C++++++++++++++++++++++++++++++++++++++++++++++++++++++++++
C
C Determine eigenvalues and eigenvectors of a symmetric,
C tridiagonal matrix.
```

```
C
C To call:     CALL TQL2(NM,N,D,E,Z,IERR)     where
C
C NM = row dimension of Z;
C N = order of matrix Z;
C D = vector of dim. N containing, on output, eigenvalues;
C E = working vector of dim. at least N-1;
C Z = matrix of dims. NM by N containing, on output, eigenvectors;
C IERR = error, normally 0, but 1 if no convergence.
C
C Normally the call to TQL2 will be preceded by a call to TRED2 in
C order to set up the tridiagonal matrix.
C
C Algorithm used: QL method of Bowdler et al., Num. Math. 11,
C 293-306, 1968.
C
C Reference: Smith et al., Matrix Eigensystem Routines - EISPACK
C Guide, Lecture Notes in Computer Science 6, Springer-Verlag,
C 1976, pp. 468-474.
C
C-------------------------------------------------------------------
      SUBROUTINE TQL2(NM,N,D,E,Z,IERR)
C
      REAL    D(N), E(N), Z(NM,N)
      DATA    EPS/1.E-12/
C
      IERR = 0
      IF (N.EQ.1) GOTO 1001
      DO 100 I = 2, N
         E(I-1) = E(I)
  100 CONTINUE
      F = 0.0
      B = 0.0
      E(N) = 0.0
C
      DO 240 L = 1, N
         J = 0
         H = EPS * (ABS(D(L)) + ABS(E(L)))
         IF (B.LT.H) B = H
C        Look for small sub-diagonal element.
         DO 110 M = L, N
            IF (ABS(E(M)).LE.B) GOTO 120
C           E(N) is always 0, so there is no exit through the
C           bottom of the loop.
  110    CONTINUE
  120    IF (M.EQ.L) GOTO 220
  130    IF (J.EQ.30) GOTO 1000
         J = J + 1
C        Form shift.
         L1 = L + 1
         G = D(L)
         P = (D(L1)-G)/(2.0*E(L))
         R = SQRT(P*P+1.0)
```

4.4. SOFTWARE AND SAMPLE IMPLEMENTATION

```
              D(L) = E(L)/(P+SIGN(R,P))
              H = G-D(L)
C
              DO 140 I = L1, N
                 D(I) = D(I) - H
   140        CONTINUE
C
              F = F + H
C             QL transformation.
              P = D(M)
              C = 1.0
              S = 0.0
              MML = M - L
C
              DO 200 II = 1, MML
                 I = M - II
                 G = C * E(I)
                 H = C * P
                 IF (ABS(P).LT.ABS(E(I))) GOTO 150
                 C = E(I)/P
                 R = SQRT(C*C+1.0)
                 E(I+1) = S * P * R
                 S = C/R
                 C = 1.0/R
                 GOTO 160
   150           C = P/E(I)
                 R = SQRT(C*C+1.0)
                 E(I+1) = S * E(I) * R
                 S = 1.0/R
                 C = C * S
   160           P = C * D(I) - S * G
                 D(I+1) = H + S * (C * G + S * D(I))
C                Form vector.
                 DO 180 K = 1, N
                    H = Z(K,I+1)
                    Z(K,I+1) = S * Z(K,I) + C * H
                    Z(K,I) = C * Z(K,I) - S * H
   180           CONTINUE
   200        CONTINUE
              E(L) = S * P
              D(L) = C * P
              IF (ABS(E(L)).GT.B) GOTO 130
   220        D(L) = D(L) + F
   240     CONTINUE
C
C          Order eigenvectors and eigenvalues.
           DO 300 II = 2, N
              I = II - 1
              K = I
```

CHAPTER 4. DISCRIMINANT ANALYSIS

```
              P = D(I)
              DO 260 J = II, N
                 IF (D(J).GE.P) GOTO 260
                 K = J
                 P = D(J)
  260         CONTINUE
              IF (K.EQ.I) GOTO 300
              D(K) = D(I)
              D(I) = P
              DO 280 J = 1, N
                 P = Z(J,I)
                 Z(J,I) = Z(J,K)
                 Z(J,K) = P
  280         CONTINUE
  300      CONTINUE
C
           GOTO 1001
C     Set error - no convergence after 30 iterations.
 1000      IERR = 1
 1001      RETURN
           END
C++++++++++++++++++++++++++++++++++++++++++++++++++++++++
C
C  Output eigenvalues in order of decreasing value.
C
C----------------------------------------------------------
           SUBROUTINE OUTEVL(N,M,NG,VALS)
           DIMENSION      VALS(NG)
C
           TOT = 0.0
           DO 100 K = 2, NG
              TOT = TOT + VALS(K)
  100      CONTINUE
C
           WRITE (6,1000)
           CUM = 0.0
           K = NG + 1
           WRITE (6,1010)
           WRITE (6,1020)
  200      CONTINUE
           K = K - 1
           CUM = CUM + VALS(K)
           VPC = VALS(K) * 100.0 / TOT
           VCPC = CUM * 100.0 / TOT
           WRITE (6,1030) VALS(K),VPC,VCPC
           IF (K.GT.2) GOTO 200
C
           RETURN
 1000 FORMAT(1H0,'EIGENVALUES FOLLOW.',/)
```

4.4. SOFTWARE AND SAMPLE IMPLEMENTATION

```fortran
 1010   FORMAT
     X(' Eigenvalues        As Percentages    Cumul. Percentages')
 1020   FORMAT
     X(' -----------        --------------    ------------------')
 1030   FORMAT(F10.4,9X,F10.4,10X,F10.4)
        END
C+++++++++++++++++++++++++++++++++++++++++++++++++++++++++++++
C
C  Output FIRST SEVEN eigenvectors associated with eigenvalues in
C  decreasing order.
C
C---------------------------------------------------------------
        SUBROUTINE OUTEVC(N1,N2,N3,VECS,N4)
        DIMENSION     VECS(N3,N3)
C
        NUM = MINO(N4,7)
        WRITE (6,1000)
        WRITE (6,1010)
        WRITE (6,1020)
        DO 100 K1 = 1, N3
        WRITE (6,1030) K1,(VECS(K1,K2),K2=1,NUM)
  100   CONTINUE
C
        RETURN
 1000   FORMAT(1H0,'EIGENVECTORS FOLLOW.',/)
 1010   FORMAT
     X  (' VBLE.   EV-1    EV-2    EV-3    EV-4    EV-5    EV-6
     X   EV-7')
 1020   FORMAT
     X  (' ------  ------  ------  ------  ------  ------  ------
     X------')
 1030   FORMAT(I5,2X,7F8.4)
        END
C+++++++++++++++++++++++++++++++++++++++++++++++++++++++++++++
C
C  Output projections on discriminant factors.
C
C---------------------------------------------------------------
        SUBROUTINE OUTPRX(N,M,NG,PRJN)
        REAL    PRJN(N,M)
C
        NUM = MINO(N,M,NG,7)
        WRITE (6,1000)
        WRITE (6,1010)
        WRITE (6,1020)
        DO 100 K = 1, N
            WRITE (6,1030) K,(PRJN(K,J),J=1,NUM-1)
  100   CONTINUE
C
```

```
      1000    FORMAT(1H0,'PROJECTIONS OF ROW-POINTS FOLLOW.',/)
      1010    FORMAT
     X        ('  OBJECT   PROJ-1   PROJ-2   PROJ-3   PROJ-4   PROJ-5   PROJ-6
     X        PROJ-7')
      1020    FORMAT
     X        ('  ------   ------   ------   ------   ------   ------   ------
     X        ------')
      1030    FORMAT(I5,2X,7F8.4)
              RETURN
              END
C++++++++++++++++++++++++++++++++++++++++++++++++++++++++++++
C
C  Output projections of column points on up to first 7
C  discriminant axes.
C
C-------------------------------------------------------------
              SUBROUTINE OUTPRY(N,M,NG,PRJNS)
              REAL     PRJNS(NG,NG)
C
              NUM = MIN0(N,M,MG,7)
              WRITE (6,1000)
              WRITE (6,1010)
              WRITE (6,1020)
              DO 100 K = 1, M
                 WRITE (6,1030) K,(PRJNS(K,J),J=1,NUM)
       100    CONTINUE
C
      1000    FORMAT(1H0,'PROJECTIONS OF COLUMN-POINTS FOLLOW.',/)
      1010    FORMAT
     X        ('  VBLE.    PROJ-1   PROJ-2   PROJ-3   PROJ-4   PROJ-5   PROJ-6
     X        PROJ-7')
      1020    FORMAT
     X        ('  ------   ------   ------   ------   ------   ------   ------
     X        ------')
      1030    FORMAT(I5,2X,7F8.4)
              RETURN
              END
C++++++++++++++++++++++++++++++++++++++++++++++++++++++++++++
C
C  Form projections of row-points on (up to) first 7 factors.
C
C-------------------------------------------------------------
              SUBROUTINE PROJX(N,M,NG,DATA,MEAN,EVEC,VEC,TOTINV)
              REAL     DATA(N,M), EVEC(M,M), VEC(M), TOTINV(M,M), MEAN(M)
C
              NUM = MIN0(N,M,NG,7)
              DO 300 K = 1, N
                 DO 50 L = 1, M
                    VEC(L) = DATA(K,L)
```

4.4. SOFTWARE AND SAMPLE IMPLEMENTATION

```
       50       CONTINUE
                DO 200 I = 1, NUM
                   DATA(K,I) = 0.0
                   DO 100 J1 = 1, M
C                     DO 75 J2 = 1, M
                         DATA(K,I) = DATA(K,I) + (VEC(J1) - MEAN(J1))*
     X                               EVEC(J1,I)
       75            CONTINUE
      100         CONTINUE
      200      CONTINUE
      300   CONTINUE
C
          RETURN
          END
C++++++++++++++++++++++++++++++++++++++++++++++++++++++++++++++
C
C  Determine projections of column points on (up to) 7 factors.
C
C----------------------------------------------------------------
          SUBROUTINE PROJY(N,M,NG,EVALS,A,Z,VEC)
          REAL    EVALS(M), A(M,M), Z(M,M), VEC(M)
C
          NUM = MIN0(N,M,NG,7)
          DO 300 J1 = 1, M
             DO 50 L = 1, M
                VEC(L) = A(J1,L)
       50    CONTINUE
             DO 200 J2 = 1, NUM
                A(J1,J2) = 0.0
                DO 100 J3 = 1, M
                   A(J1,J2) = A(J1,J2) + VEC(J3)*Z(J3,M-J2+1)
      100       CONTINUE
                IF (EVALS(M-J2+1).GT.0.0) A(J1,J2) =
     X             A(J1,J2)/SQRT(EVALS(M-J2+1))
                IF (EVALS(M-J2+1).EQ.0.0) A(J1,J2) = 0.0
      200    CONTINUE
      300 CONTINUE
C
          RETURN
          END
```

4.4.3 Program Listing: K–NNs Discriminant Analysis

```
C++++++++++++++++++++++++++++++++++++++++++++++++++++++
C
C   Carry out a K-NN DISCRIMINANT ANALYSIS,
C   with two groups defined in a training set, and with
C   assignment of members of a test set.
C
C   Parameters:
C
C   TRAIN(N,M)      training set, where first N1 rows relate to the
C                   first group, and the next N2 rows to the second
C                   group. Must have N1 + N2 = N.
C   TEST(N3,M)      test set;
C   K               number of nearest neighbours to consider;
C   KLIST(K), DK(K), KPOP(K)  are used for storing the K NNs,
C                   their distances to the object under consider-
C                   ation, and the group to which the NNs belong.
C
C-----------------------------------------------------------------
        SUBROUTINE KNN(TRAIN,N1,N2,N,TEST,N3,M,K,KLIST,DK,KPOP)
        DIMENSION TRAIN(N,M),TEST(N3,M)
        DIMENSION KLIST(K),DK(K),KPOP(K)
C
        WRITE(6,22)
        WRITE(6,33)
C
        DO 90 I = 1, N3
        DO 10 IX = 1, K
        KLIST(IX) = 0
        DK(IX) = 1.E+15
        KPOP(IX) = 0
   10   CONTINUE
        IND = 0
                DO 38 I2 = 1, N1
                CALL DIST(I2,TRAIN,I,TEST,N,N3,M,D)
                IF (IND.LE.0) GOTO 35
                   DO 30 IK = 1, K
                      IF (D.GE.DK(IK)) GOTO 28
                         IF (IK.GE.K) GOTO 25
                            DO 20 IK2 = K, IK+1, -1
                               DK(IK2) = DK(IK2-1)
                               KLIST(IK2) = KLIST(IK2-1)
                               KPOP(IK2) = KPOP(IK2-1)
   20                       CONTINUE
   25                    CONTINUE
                         DK(IK) = D
                         KLIST(IK) = I2
                         KPOP(IK) = 1
                         GOTO 36
```

4.4. SOFTWARE AND SAMPLE IMPLEMENTATION

```
   28            CONTINUE
   30            CONTINUE
                 GOTO 36
   35            CONTINUE
                    IND = IND + 1
                    DK(IND) = D
                    KLIST(IND) = I2
                    KPOP(IND) = 1
   36            CONTINUE
   38            CONTINUE
C
C
C
             DO 68 I2 = N1+1, N
             CALL DIST(I2,TRAIN,I,TEST,N,N3,M,D)
             IF (IND.LE.0) GOTO 65
                DO 60 IK = 1, K
                   IF (D.GE.DK(IK)) GOTO 58
                      IF (IK.GE.K) GOTO 55
                         DO 50 IK2 = K, IK+1, -1
                            DK(IK2) = DK(IK2-1)
                            KLIST(IK2) = KLIST(IK2-1)
                            KPOP(IK2) = KPOP(IK2-1)
   50                    CONTINUE
   55                 CONTINUE
                      DK(IK) = D
                      KLIST(IK) = I2
                      KPOP(IK) = 2
                      GOTO 66
   58             CONTINUE
   60          CONTINUE
               GOTO 66
   65          CONTINUE
                  IND = IND + 1
                  DK(IND) = D
                  KLIST(IND) = I2
                  KPOP(IND) = 2
   66          CONTINUE
   68       CONTINUE
C
         NUM1 = 0
         NUM2 = 0
         DO 80 IX = 1, K
            IF (KPOP(IX).EQ.1) NUM1 = NUM1 + 1
            IF (KPOP(IX).EQ.2) NUM2 = NUM2 + 1
   80    CONTINUE
C     (Error check:)
         IF ((NUM1+NUM2).EQ.K) GOTO 85
            WRITE (6,600)
```

```
            STOP
   85   CONTINUE
        IF (NUM1.GT.NUM2) WRITE (6,500) I,FLOAT(NUM1)*100./FLOAT(K)
        IF (NUM2.GT.NUM1) WRITE (6,525) I,FLOAT(NUM2)*100./FLOAT(K)
        IF (NUM1.EQ.NUM2) WRITE (6,550) I,FLOAT(NUM1)*100./FLOAT(K)
   90   CONTINUE
C
        RETURN
   22   FORMAT(' Object   -->   group with probability')
   33   FORMAT(/)
  500   FORMAT(I6,6X,'     1    ',F8.2,'%')
  525   FORMAT(I6,6X,'     2    ',F8.2,'%')
  550   FORMAT(I6,6X,' 1 or 2  ',F8.2,'%   (equiprobable).')
  600   FORMAT(' The total of assignments to gp. 1 and to gp. 2',
       X        'does not equal K; check pgm. listing; aborting.')
        END
C-----------------------------------------------------------------------
        SUBROUTINE DIST(I,ARR1,J,ARR2,NR1,NR2,NC,D)
        DIMENSION ARR1(NR1,NC),ARR2(NR2,NC)
C
        D = 0.0
        DO 10 K = 1, NC
        D = D + (ARR1(I,K)-ARR2(J,K))**2
   10   CONTINUE
C
        RETURN
        END
```

4.4. SOFTWARE AND SAMPLE IMPLEMENTATION

4.4.4 Input Data

The input data was based on that used for Principal Components Analysis. The class assignments used were as follows. For LDA, the first seven objects (rows) constituted group 1, and the remaining objects constituted group 2. For MDA, three groups were used: objects 1 to 4, objects 5 to 14, and objects 15 to 18. For KNN, the first 10 objects were taken as group 1 and the remainder as group 2. The training set and the test set in the case of this last method were the same.

4.4.5 Sample Output: Linear Discriminant Analysis

PROJECTIONS OF ROW-POINTS FOLLOW.

OBJECT	PROJN
1	2.8601
2	-0.0343
3	3.1341
4	4.9368
5	1.0754
6	2.0656
7	2.2036
8	-2.4989
9	1.3870
10	-1.1907
11	-3.6095
12	-2.0898
13	-2.8095
14	-2.0526
15	0.4962
16	-0.7544
17	-5.3203
18	-18.5029

4.4.6 Sample Output: Multiple Discriminant Analysis

EIGENVALUES FOLLOW.

Eigenvalues	As Percentages	Cumul. Percentages
1.6564	72.0512	72.0512
0.6425	27.9488	100.0000

EIGENVECTORS (in the group space) FOLLOW.

VBLE.	EV-1	EV-2	EV-3	EV-4	EV-5	EV-6	EV-7
1	0.6370	0.7188					
2	0.5233	-0.6685					
3	0.5660	-0.1908					

EIGENVECTORS (in the parameter space) FOLLOW.

VBLE.	EV-1	EV-2	EV-3	EV-4	EV-5	EV-6	EV-7
1	-3.1656	-3.9568					
2	1.4853	3.8207					
3	5.7957	3.6735					
4	-4.6384	-2.2279					
5	1.5489	-2.4221					
6	-1.0008	1.1459					
7	-0.1667	-0.0222					

4.4. SOFTWARE AND SAMPLE IMPLEMENTATION

```
 8   -0.0372  0.0373
 9    0.2003  0.0713
10    0.1152 -0.1726
11   -0.1829 -0.1045
12   -0.5876 -0.5799
13    0.5090  0.2943
14   -0.0520  0.2415
15    0.0722 -0.0647
16   -0.0365 -0.1560
```

PROJECTIONS OF ROW-POINTS FOLLOW.

```
OBJECT  PROJ-1  PROJ-2  PROJ-3  PROJ-4  PROJ-5  PROJ-6  PROJ-7
------  ------  ------  ------  ------  ------  ------  ------
   1    0.2340  2.2248
   2    0.2246  3.0197
   3   -0.0729  1.1148
   4    0.4106  1.7577
   5    0.6678  0.2472
   6    0.0369  0.1120
   7    0.1016  0.0550
   8    0.3308 -0.7271
   9    1.0187  0.4180
  10    0.3271 -0.6839
  11   -0.2767 -0.6126
  12    0.8632 -0.1824
  13   -0.1438 -1.0527
  14    0.6162 -0.0902
  15   -0.9246 -0.4047
  16   -1.9487 -1.0058
  17   -1.2916 -1.4706
  18   -0.1734 -2.7195
```

4.4.7 Sample Output: K–NNs Discriminant Analysis

```
Object  -->  group with probability
   1          1       100.00%
   2          1       100.00%
   3          1       100.00%
   4          1       100.00%
   5          1       100.00%
   6          1       100.00%
   7          1       100.00%
   8          1       100.00%
   9          1       100.00%
  10          1        66.67%
  11          2        66.67%
  12          2       100.00%
  13          2       100.00%
  14          2       100.00%
  15          2       100.00%
  16          2       100.00%
  17          2       100.00%
  18          2       100.00%
```

Chapter 5
Other Methods

5.1 The Problems

Principal Components Analysis (PCA), among whose objectives are dimensionality reduction and the display of data, assumes points in the usual Euclidean space as input. For other types of input data, alternative methods exist. Such other methods may have somewhat different objectives, which may be more relevant for the given type of input.

Correspondence Analysis is particularly suitable for arrays of frequencies or for data in complete disjunctive form (cf. Chapter 1). It may be described as a PCA in a different metric (the χ^2 metric replaces the usual Euclidean metric). Mathematically, it differs from PCA also in that points in multidimensional space are considered to have a mass (or weight) associated with them, at their given locations. The percentage *inertia* explained by axes takes the place of the percentage variance of PCA, — and in the former case the values can be so small that such a figure of merit assumes less importance than in the case of PCA. Correspondence Analysis is a technique in which it is a good deal more difficult to interpret results, but it considerably expands the scope of a PCA–type analysis in its ability to handle a wide range of data.

Principal Coordinates Analysis is very similar to PCA. The problem here is that rather than the usual objects × variables array, we are given an objects × objects distance matrix. A minimal amount of alteration to the approach adopted in PCA allows this type of input data to be handled.

In Canonical Correlation Analysis, the third technique to be looked at in this Chapter, we are given a set of objects (rows) crossed by two distinct sets of variables (columns). The objective is to examine the relationship between the two sets of characterisations of the same object–population. A pair of best–fitting axes are derived in the spaces of the two sets of variables, such that in addition these two axes are optimally correlated. Successive axes are subsequently obtained.

Canonical Correlation Analysis is difficult to use in practice when the two sets of variables are not highly related.

Regression Analysis, the final topic of this Chapter, is widely dealt with elsewhere in the physical and statistical literature (see in particular Lutz, 1983). For completeness in the range of important multivariate methods studied, we introduce it and discuss a few points of relevance to astronomy. Relevant bibliographic references in astronomy in this Chapter primarily relate to regression.

5.2 Correspondence Analysis

5.2.1 Introduction

Correspondence Analysis (CA) is not unlike PCA in its underlying geometrical bases, and the description to follow will adopt a similar perspective to that used in Chapter 2. While PCA is particularly suitable for quantitative data, CA is recommendable for the following types of input data, which will subsequently be looked at more closely: frequencies, contingency tables, probabilities, categorical data, and mixed qualitative/categorical data.

In the case of *frequencies* (i.e. the ij^{th} table entry indicates the frequency of occurrence of attribute j for object i) the row and column "profiles" are of interest. That is to say, the relative magnitudes are of importance. Use of a weighted Euclidean distance, termed the χ^2 distance, gives a zero distance for example to the following 5-coordinate vectors which have identical *profiles* of values: (2,7,0,3,1) and (8,28,0,12,4). Probability type values can be constructed here by dividing each value in the vectors by the sum of the respective vector values.

A particular type of frequency of occurrence data is the *contingency table*, — a table crossing (usually, two) sets of characteristics of the population under study. As an example, an $n \times m$ contingency table might give frequencies of the existence of n different metals in stars of m different ages. CA allows the study of the two sets of variables which constitute the rows and columns of the contingency table. In its usual variant, PCA would privilege either the rows or the columns by standardizing (cf. Section 2.2.5): if, however, we are dealing with a contingency table, both rows and columns are equally interesting. The "standardizing" inherent in CA (a consequence of the χ^2 distance) treats rows and columns in an identical manner. One byproduct is that the row and column projections in the new space may both be plotted on the same output graphic presentations (— the lack of an analogous direct relationship between row projections and column projections in PCA precludes doing this in the latter technique).

Categorical data may be coded by the "scoring" of 1 (presence) or 0 (absence) for each of the possible categories. Such coding leads to *complete disjunctive coding*, as seen in Chapter 1. It will be discussed below how CA of an array of

5.2. CORRESPONDENCE ANALYSIS

such complete disjunctive data is referred to as Multiple Correspondence Analysis (MCA), and how such a coding of categorical data is, in fact, closely related to contingency table type data.

Dealing with a complex astronomical catalogue may well give rise in practice to a mixture of quantitative (real valued) and qualitative data. One possibility for the analysis of such data is to "discretize" the quantitative values, and treat them thereafter as categorical. In this way a set of variables — many more than the initially given set of variables — which is homogeneous, is analysed.

CA is described initially below with reference to frequency or probability type data as input. We will then look at how precisely the same method is also used for complete disjunctive data.

5.2.2 Properties of Correspondence Analysis

From the initial frequencies data matrix, a set of probability data, x_{ij}, is defined by dividing each value by the grand total of all elements in the matrix. In CA, each row (or column) point is considered to have an associated weight. The weight of the ith row point is given by $x_i = \sum_j x_{ij}$ and the weight of the jth column point is given by $x_j = \sum_i x_{ij}$. We consider the row points to have coordinates x_{ij}/x_i, thus allowing points of the same *profile* to be identical (i.e. superimposed). The following weighted Euclidean distance, the χ^2 distance, is then used between row points:

$$d^2(i,k) = \sum_j \frac{1}{x_j} \left(\frac{x_{ij}}{x_i} - \frac{x_{kj}}{x_k} \right)^2$$

and an analogous distance is used between column points.

Table 5.1 summarizes the situation in the dual spaces. Item 4 indicates that the Euclidean distance between points of coordinates x_{ij}/x_i and x_{ij}/x_j is not the ordinary Euclidean distance but is instead with respect to the specified weights. Item 5 indicates that the inertia rather than the variance will be studied, i.e. that the masses of the points are incorporated into the criterion to be optimised.

The mean row point is given by the weighted average of all row points:

$$\sum_i x_i \frac{x_{ij}}{x_i} = x_j$$

for $j = 1, 2, \ldots, m$. Similarly the mean column profile has i^{th} coordinate x_i.

Space $\mathbb{R}^m$:

1. n row points, each of m coordinates.

2. The jth coordinate is x_{ij}/x_i.

3. The mass of point i is x_i.

4. The χ^2 distance between i and k is:
$d^2(i,k) = \sum_j \frac{1}{x_j}(\frac{x_{ij}}{x_i} - \frac{x_{kj}}{x_k})^2$.
Hence this is a Euclidean distance with respect to the weighting $1/x_j$ (for all j).

5. The criterion to be optimised: the weighted sum of squares of projections, where the weighting is given by x_i (for all i).

Space $\mathbb{R}^n$:

1. m column points, each of n coordinates.

2. The ith coordinate is x_{ij}/x_j.

3. The mass of point j is x_j.

4. The χ^2 distance between column points g and j is:
$d^2(g,j) = \sum_i \frac{1}{x_i}(\frac{x_{ig}}{x_g} - \frac{x_{ij}}{x_j})^2$.
Hence this is a Euclidean distance with respect to the weighting $1/x_i$ (for all i).

5. The criterion to be optimised: the weighted sum of squares of projections, where the weighting is given by x_j (for all j).

Table 5.1: Properties of spaces $\mathbb{R}^m$ and $\mathbb{R}^n$ in Correspondence Analysis.

5.2. CORRESPONDENCE ANALYSIS

5.2.3 The Basic Method

As in the case of PCA, we first consider the projections of the n profiles in $\mathbb{R}^m$ onto an axis, $\mathbf{u}$. This is given by

$$\sum_j \frac{x_{ij}}{x_i} \frac{1}{x_j} u_j$$

for all i (note how the scalar product, used here, is closely related to the definition of distance — item 4 in Table 5.1). Let the above, for convenience, be denoted by w_i.

The weighted sum of projections uses weights x_i (i.e. the row masses), since the inertia of projections is to be maximized. Hence the quantity to be maximized is

$$\sum_i x_i w_i^2.$$

subject to the vector $\mathbf{u}$ being of unit length (this, as in PCA, is required since otherwise vector $\mathbf{u}$ could be taken as unboundedly large):

$$\sum_j \frac{1}{x_j} u_j^2 = 1.$$

It may then be verified using Lagrangian multipliers that optimal $\mathbf{u}$ is an eigenvector of the matrix of dimensions $m \times m$ whose $(j,k)^{th}$ term is

$$\sum_i \frac{x_{ij} x_{ik}}{x_i x_k}$$

where $1 \le j, k \le m$. (Note that this matrix is not symmetric, and that a related symmetric matrix must be constructed for eigenreduction: we will not detail this here.) The associated eigenvalue, λ, indicates the importance of the best fitting axis, or eigenvalue: it may be expressed as the *percentage of inertia explained* relative to subsequent, less good fitting, axes.

The results of a CA are centred (x_j and x_i are the jth and i^{th} coordinates — the average profiles — of the origin of the output graphic representations). The first eigenvalue resulting from CA is a trivial one, of value 1; the associated eigenvector is a vector of 1s (Lebart et al., 1984; Volle, 1981).

5.2.4 Axes and Factors

In the previous section it has been seen that projections of points onto axis $\mathbf{u}$ were with respect to the $1/x_i$ weighted Euclidean metric. This makes interpreting projections very difficult from a human/visual point of view, and so it is more natural to present results in such a way that projections can be simply appreciated. Therefore *factors* are defined, such that the projections of row vectors onto factor ϕ associated with axis **u** are given by

$$\sum_j \frac{x_{ij}}{x_i} \phi_j$$

for all i. Taking

$$\phi_j = \frac{1}{x_j} u_j$$

ensures this and projections onto ϕ are with respect to the ordinary (unweighted) Euclidean distance.

An analogous set of relationships hold in $\mathbb{R}^n$ where the best fitting axis, **v**, is searched for. A simple mathematical relationship holds between **u** and **v**, and between ϕ and ψ (the latter being the factor associated with eigenvector **v**):

$$\sqrt{\lambda}\psi_i = \sum_j \frac{x_{ij}}{x_i} \phi_j$$

$$\sqrt{\lambda}\phi_j = \sum_i \frac{x_{ij}}{x_j} \psi_i.$$

These are termed *transition formulas*. Axes **u** and **v**, and factors ϕ and ψ, are associated with eigenvalue λ and best fitting higher–dimensional subspaces are associated with decreasing values of λ, determined in the diagonalization.

The transition formulas allow *supplementary rows* or columns to be projected into either space. If ξ_j is the jth element of a supplementary row, with mass ξ, then a factor loading is simply obtained subsequent to the CA:

$$\psi_i = \frac{1}{\sqrt{\lambda}} \sum_j \frac{\xi_j}{\xi} \phi_j.$$

A similar formula holds for supplementary columns. Such supplementary elements are therefore "passive" and are incorporated into the CA results subsequent to the CA being carried out.

5.2.5 Multiple Correspondence Analysis

When the input data is in *complete disjunctive form*, CA is termed Multiple CA (MCA). Complete disjunctive form is a form of coding where the response categories, or modalities, of an attribute have one and only one non–zero response (see Figure 5.1a). Ordinarily CA is used for the analysis of contingency tables: such a table may be derived from a table in complete disjunctive form by taking

5.2. CORRESPONDENCE ANALYSIS

	Type			Age			Properties			
T1	T2	T3	A1	A2	A3	P1	P2	P3	P4	P5
1	0	0	0	1	0	0	0	0	0	1
0	1	0	0	0	1	0	0	0	0	1
1	0	0	0	0	1	0	0	1	0	0
1	0	0	0	1	0	0	0	1	0	0
1	0	0	1	0	0	1	0	0	0	0
0	0	1	0	1	0	1	0	0	0	0
0	0	1	1	0	0	0	0	0	0	1

(a) Table in complete disjunctive form.

	T1	T2	T3	A1	A2	A3	P1	P2	P3	P4	P5
T1	4	0	0	1	2	1	1	0	2	0	1
T2	0	1	0	0	0	1	0	0	0	0	1
T3	0	0	2	1	1	0	1	0	0	0	1
A1	1	0	1	2	0	0	1	0	0	0	1
A2	2	0	1	0	3	0	1	0	1	0	1
A3	1	1	0	0	0	2	0	0	1	0	1
P1	1	0	1	1	1	0	2	0	0	0	0
P2	0	0	0	0	0	0	0	0	0	0	0
P3	2	0	0	0	1	1	0	0	2	0	0
P4	0	0	0	0	0	0	0	0	0	0	0
P5	1	1	1	1	1	1	0	0	0	0	3

(b) Burt table.

Notes:

- Attributes: Type, Age, Properties.

- Modalities: T1, T2, ..., P5.

- Row sums of table in complete disjunctive form are constant, and equal the number of attributes.

- Each attribute × attribute submatrix of the Burt table (e.g. Ages × Ages) is necessarily diagonal, with column totals of the table in complete disjunctive form making up the diagonal values.

Figure 5.1: Table in complete disjunctive form and associated Burt table.

the matrix product between its transpose and itself. The symmetric table obtained in this way is referred to as a *Burt table*. CA of either table gives similar results, only the eigenvalues differing (see Lebart *et al.*, 1984; or Volle, 1981).

A few features of the analysis of tables in complete disjunctive form will be mentioned.

- The modalities (or response categories) of each attribute in MCA have their centre of gravity at the origin.

- The number of nonzero eigenvalues found is less than or equal to the total number of modalities less the total number of attributes.

- Due to this large dimensionality of the space being analyzed, it is not surprising that eigenvalues tend to be very small in MCA. It is not unusual to find that the first few factors can be usefully interpreted and yet account for only a few percent of the total inertia.

The principal steps in interpreting the output of MCA, as in CA, are similar to the interpreting of PCA output.

- The Burt table is scanned for significantly high frequencies of co–occurrence.

- The axes are interpreted in order of decreasing importance using the modalities which contribute most, in terms of inertia, to the axes (i.e. mass times projected distance squared). The projection coordinates serve to indicate how far the modality can be assessed relative to the axis.

- The planar graphic representations (projections of row and column points in the plane formed by factors 1 and 2, and by other pairs of factors) are examined.

- The interrelationships between modalities, relative to the axes, are examined, and substantive conclusions are drawn.

It may be noted that in the variant of Correspondence Analysis looked at in this section, the row–points are of constant weight. This allows, quite generally, user intervention in the weighting of rows relative to columns. In our experience, we have often obtained very similar results for a Principal Components Analysis with the usual standardization to zero mean and unit standard deviation, on the one hand; and on the other, a Correspondence Analysis with twice the number of columns as the matrix analyzed by PCA such that for each column j we also have a column j' with value $x_{ij'} = \max_k x_{ik} - x_{ij}$. This is referred to as *doubling* the data.

Some typical output configurations can arise, the most well known being the "horseshoe" shaped curve associated with pronounced linearity in the data. Figure 5.3 gives an example of the type of doubled data for which this pattern arises. It

5.3 Principal Coordinates Analysis

5.3.1 Description

Principal Coordinates Analysis has also been referred to as *Classical Multidimensional Scaling* and *metric scaling*, and has been associated with the names of Torgerson and Gower. It takes distances as input and produces coordinate values as output. In this, it has been described as producing something from nothing.

It may be useful to review a possible situation where distance input is readily available whereas the raw data are not. Consider a catalogue or database information which contains attributes which are a mixture of quantitative and qualitative (categorical) types. Correspondence Analysis offers one approach to the analysis of such data, by recoding all data in a qualitative form, and using the complete disjunctive form of coding seen above. An alternative is to use a distance for mixed data derived from the Gower coefficient described in Chapter 1, and then to use Principal Coordinates Analysis on such data.

We now describe the principles on which Principal Coordinates Analysis is based, and how it is implemented. Consider the initial data matrix, X, of dimensions $n \times m$, and the "sums of squares and cross products" matrix of the rows:

$$A = XX'$$

$$a_{ik} = \sum_j x_{ij} x_{kj}.$$

If d_{ik} is the Euclidean distance between objects i and k (using row–vectors i and k of matrix X) we have that:

$$d_{ik}^2 = \sum_j (x_{ij} - x_{kj})^2$$

$$= \sum_j x_{ij}^2 + \sum_j x_{kj}^2 - 2 \sum_j x_{ij} x_{kj}$$

$$= a_{ii} + a_{kk} - 2 a_{ik}. \qquad (5.1)$$

In Principal Coordinates Analysis, we are given the distances and we want to obtain X. We will assume that the columns of this matrix are centred, i.e.

$$\sum_i x_{ij} = 0.$$

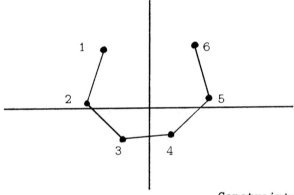

Figure 5.2: Horseshoe pattern in principal plane of Correspondence Analysis.

5.3. PRINCIPAL COORDINATES ANALYSIS

It will now be shown that matrix A can be constructed from the distances using the following formula:

$$a_{ik} = -\frac{1}{2}(d_{ik}^2 - d_i^2 - d_k^2 - d^2) \tag{5.2}$$

where

$$d_i^2 = \frac{1}{n}\sum_k d_{ik}^2$$

$$d_k^2 = \frac{1}{n}\sum_i d_{ik}^2$$

$$d^2 = \frac{1}{n^2}\sum_i\sum_k d_{ik}^2.$$

This result may be proved by substituting for the distance terms (using equation 5.1), and knowing that by virtue of the centring of the row vectors of matrix X, we have

$$\sum_i a_{ik} = 0$$

(since $a_{ik} = \sum_j x_{ij}x_{kj}$; and consequently in the term $\sum_i \sum_j x_{ij}x_{kj}$ we can separate out $\sum_i x_{ij}$ which equals zero). Similarly (by virtue of the symmetry of A) we use the fact that

$$\sum_k a_{ik} = 0.$$

Having thus been given distances, we have constructed matrix $A = XX'$. We now wish to reconstruct matrix X; or, since this matrix has in fact never existed, we require some matrix X which satisfies $XX' = A$.

If matrix A is positive, symmetric and semidefinite, it will have rank $p \leq n$. We may derive p non–zero eigenvalues, $\lambda_1 \geq \lambda_2 \geq \ldots \geq \lambda_p > 0$, with corresponding eigenvectors $\mathbf{u}_1, \mathbf{u}_2, \ldots, \mathbf{u}_p$. Consider the scaled eigenvectors, defined as $\mathbf{f}_i = \sqrt{\lambda_i}\mathbf{u}_i$. Then the matrix $X = (\mathbf{f}_1, \mathbf{f}_2, \ldots, \mathbf{f}_p)$ is a possible coordinate matrix. This is proved as follows. We have, in performing the eigen–decomposition of A:

$$A\mathbf{u}_i = \lambda_i \mathbf{u}_i$$

and by requirement

$$XX'\mathbf{u}_i = \lambda_i \mathbf{u}_i.$$

In the left hand side, $X \begin{pmatrix} \mathbf{f}_1 \\ \mathbf{f}_2 \\ \cdot \\ \cdot \\ \cdot \\ \mathbf{f}_p \end{pmatrix} \mathbf{u}_i = X \begin{pmatrix} \sqrt{\lambda_1}\mathbf{u}_1 \\ \sqrt{\lambda_2}\mathbf{u}_2 \\ \cdot \\ \cdot \\ \cdot \\ \sqrt{\lambda_p}\mathbf{u}_p \end{pmatrix} \mathbf{u}_i = X \begin{pmatrix} 0 \\ 0 \\ \cdot \\ \sqrt{\lambda_i} \\ \cdot \\ 0 \end{pmatrix}$ since eigenvectors are mutually orthogonal. Continuing:

$$(\sqrt{\lambda_1}\mathbf{u}_1, \sqrt{\lambda_2}\mathbf{u}_2, \ldots, \sqrt{\lambda_p}\mathbf{u}_p) \begin{pmatrix} 0 \\ 0 \\ \vdots \\ \sqrt{\lambda_i} \\ \vdots \\ 0 \end{pmatrix} = \lambda_i \mathbf{u}_i.$$

Thus we have succeeded in constructing a matrix X, having been initially given a set of distances. The net result is very similar to PCA, — we have a series of orthogonal axes which indicate inherent dimensionality and which may be used for plotting the objects studied. We have not used any set of variables or attributes in Principal Coordinates Analysis, but on carrying out this technique we have a set of projections on principal axes which may be used as attributes.

Principal Coordinates Analysis may be programmed using a PCA program with relatively few changes. Rather than reading in the data and constructing a sums of squares and cross products (SSCP) matrix, we read in a matrix of distances and use Equation 5.2 above to form the SSCP matrix.

In practice, we might be given dissimilarities rather than distances. Then, matrix A will be symmetric and have zero values on the diagonal but will not be positive semidefinite. In this case negative eigenvalues are obtained. These are inconvenient but may often be ignored if the approximate Euclidean representation (given by the eigenvectors corresponding to positive eigenvalues) is satisfactory.

5.3.2 Multidimensional Scaling

Principal Coordinates Analysis has been referred to as *metric multidimensional scaling*. From the early 1960s onwards, *non–metric multidimensional scaling* was developed and became widely used especially in the areas of psychology and marketing. Given dissimilarities, δ_{ij}, it estimates (using iterative optimization) Euclidean distances, d_{ij}, such that rank orders are preserved as far as possible:

$$\delta_{ij} \leq \delta_{kl} \Rightarrow d_{ij} \leq d_{kl}.$$

A figure of merit for the Euclidean configuration in a space of some given dimension is the "stress", a definition of which is

$$\sum_{i>j}(d_{ij} - \delta_{ij})^2 / \sum_{i>j} d_{ij}^2.$$

A space of low dimension is sought, coupled with a low value of the stress criterion. Non–metric multidimensional scaling is less restrictive in its assumptions about input data compared to Principal Coordinates Analysis, but it may require greater computation time. Further reading is to be found in Gordon (1981) and Kruskal and Wish (1978).

5.4 Canonical Correlation Analysis

In Canonical Correlation (or Variate) Analysis, we are given n objects crossed by two sets of p and q variables, respectively. The $n \times p$ matrix, X, and the $n \times q$ matrix, Y, together give the raw data matrix Z of dimensions $n \times (p+q)$. As examples of problems suitable for study using Canonical Correlation Analysis, consider n objects with attributes defined by two different types of instrument, by two different sets of observers, using two different photometric systems, and so on.

Two sets of linear combinations of row i's coordinates are defined, relative to X and relative to Y:

$$u_i = \sum_{j=1}^{p} a_j x_{ij} \qquad \mathbf{u} = X\mathbf{a}$$

$$v_i = \sum_{j=1}^{q} b_j y_{ij} \qquad \mathbf{v} = Y\mathbf{b}$$

where $\mathbf{a}$ and $\mathbf{b}$ are vectors of constants of dimensions, respectively, $p \times 1$ and $q \times 1$.

Canonical Correlation Analysis seeks two sets of linear combinations, $\mathbf{u}$ and $\mathbf{v}$, such that these are as highly correlated as possible. As in PCA (Chapter 2), the restriction is imposed that $\mathbf{u}$ and $\mathbf{v}$ have unit variance; and we proceed to an optimization, using Lagrangian multipliers, which results in an eigenvalue equation. Successive pairs of correlated linear combinations are determined.

The eigenvalues are referred to as *canonical correlation coefficients*. If the first two coefficients are close to 1, there is a strong resemblance between the two sets of variables, as measured on the first two canonical variates. For a planar representation of the variables, we may select the canonical variates of either group of variables, and project the other group onto this plane. As in PCA we will seek to interpret the axes (points which are projected towards the extremities may offer greatest help for doing this); and to look for clusters of variables, if possible leading to indications of close association between the variables of the two sets used.

Canonical Correlation Analysis is not as widely used as might be expected from the generality of the problems which it can address. This is because results are difficult to interpret, especially when canonical correlations, associating the two sets of variables, are not high.

5.5 Regression Analysis

In linear regression we are given a set of m values, $\{x_j, y_j : j = 1, \ldots m\}$ and we require a fit of the form $y = a + bx$, the *regression line of y on x*. We assume that x is fixed or controlled, i.e. that its exact value is known. The variable y is *stochastic*. It is assumed that the jth y value, y_j, is distributed about some mean $\bar{y}_j$ with standard deviation σ_j. Thus, for a given x_j, we have a certain scatter in corresponding y_j values, and the eventual linear fit will be based on $\bar{y}_j$. The distributional form of the y_j values is assumed to be Gaussian. Hence the probability of a particular value of y_j is given by

$$\frac{1}{\sqrt{2\pi}\,\sigma_j}\exp\left(-\frac{1}{2}\left(\frac{y_j-\bar{y}_j}{\sigma_j}\right)^2\right).$$

The simultaneous probability of a given set of m values is given by the product of this expression, over all j:

$$\prod_{j=1}^{m}\left(\frac{1}{\sqrt{2\pi}\,\sigma_j}\exp\left(-\frac{1}{2}\sum_{j=1}^{m}\left(\frac{y_j-\bar{y}_j}{\sigma_j}\right)^2\right)\right).$$

In order to determine $\bar{y}_j$ we employ the method of maximum likelihood: maximizing the above term implies minimizing the exponent. Hence we minimize

$$\sum_{j=1}^{m}\left(\frac{y_j-\bar{y}_j}{\sigma_j}\right)^2$$

which leads to the least squares estimate of the $\bar{y}_j$ values. (Linear regression is often introduced more directly as this least squares optimizing, with weights defined by $w_j = 1/\sigma_j^2$.) Using $\bar{y}_j = a + bx_j$, the expression to be minimized becomes

$$\sum_{j=1}^{m}\left(\frac{y_j-a-bx_j}{\sigma_j}\right)^2.$$

To solve for a and b, the partial derivatives of this term with respect to a and to b are determined, and the resulting equations solved. The expressions found in this way for a (the *intercept*) and b (the *slope* or *regression coefficient*) are given in Bevington (1969), together with approximate error bounds.

In the foregoing, a *regression of y on x* has been looked at. By supposition, there is variation in y but not in x. Some variation in x is acceptable in practice, but if both variables are considered as stochastic, this linear regression is not suitable. In the case of stochastic x and error–free y, it is of course straightforward to regress x on y.

When both x and y are stochastic, determining the best fitting straight line is studied by York (1966; see also Lybanon, 1984). In this case, analytic expressions for a and b are not feasible. Instead, given some initial estimate of the slope, b_0 (obtained, for instance, by a regression analysis ignoring the error weights on x and y), iterative optimization is used to improve the value of b.

Linear regression generalizes to *multiple regression* when a fit of the form

$$y = a_0 + a_1 x_1 + a_2 x_2 + \ldots + a_n x_n$$

is sought, for n independent variables. The coefficients, a, are called *partial regression coefficients*.

5.6 Examples and Bibliography

5.6.1 Regression in Astronomy

1. E. Antonello and M. Fracassini, "Pulsars and interstellar medium: multiple regression analysis of related parameters", *Astrophysics and Space Science*, **108**, 187–193, 1985.

2. R.L. Branham Jr., "Alternatives to least–squares", *The Astronomical Journal*, **87**, 928–937, 1982.

3. R. Buser, "A systematic investigation of multicolor photometric systems. II. The transformations between the UBV and RGU systems.", *Astronomy and Astrophysics*, **62**, 425–430, 1978.

4. C.R. Cowley and G.C.L. Aikman, "Stellar abundances from line statistics", *The Astrophysical Journal*, **242**, 684–698, 1980.

5. M. Crézé, "Influence of the accuracy of stellar distances on the estimations of kinematical parameters from radial velocities", *Astronomy and Astrophysics*, **9**, 405–409, 1970.

6. M. Crézé, "Estimation of the parameters of galactic rotation and solar motion with respect to Population I Cepheids", *Astronomy and Astrophysics*, **9**, 410–419, 1970.

7. T.J. Deeming, "The analysis of linear correlation in astronomy", *Vistas in Astronomy*, **10**, 125, 1968.

8. H. Eichhorn, "Least–squares adjustment with probabilistic constraints", *Monthly Notices of the Royal Astronomical Society*, **182**, 355–360, 1978.

9. H. Eichhorn and M. Standish, Jr., "Remarks on nonstandard least–squares problems", *The Astronomical Journal*, **86**, 156–159, 1981.

10. M. Fracassini, L.E. Pasinetti and E. Antonello, "Pulsars and interstellar medium", *Proceedings of a Course and Workshop on Plasma Astrophysics*, European Space Agency Special Publication 207, 319–321, 1984.

11. J.R. Gott III and E.L. Turner, "An extension of the galaxy covariance function to small scales", *The Astrophysical Journal*, **232**, L79–L81, 1979.

12. A. Heck, "Predictions: also an astronomical tool", in *Statistical Methods in Astronomy*, European Space Agency Special Publication 201, 1983, pp. 135–143.

    (A survey article, with many references. Other articles in this conference proceedings also use regression and fitting techniques.)

13. A. Heck and G. Mersch, "Prediction of spectral classification from photometric observations — application to the $uvby\beta$ photometry and the MK spectral classification. I. Prediction assuming a luminosity class", *Astronomy and Astrophysics*, **83**, 287–296, 1980.

    (Stepwise multiple regression and isotonic regression are used.)

14. W.H. Jefferys, "On the method of least squares", *The Astronomical Journal*, **85**, 177–181, 1980.

15. W.H. Jefferys, "On the method of least squares. II.", *The Astronomical Journal*, **86**, 149–155, 1981.

16. J.R. Kuhn, "Recovering spectral information from unevenly sampled data: two machine–efficient solutions", *The Astronomical Journal*, **87**, 196–202, 1982.

17. T.E. Lutz, "Estimation — comments on least squares and other topics", in *Statistical Methods in Astronomy*, European Space Agency Special Publication 201, 179–185, 1983.

18. M.O. Mennessier, "Corrections de précession, apex et rotation galactique estimées à partir de mouvements propres fondamentaux par une méthode de maximum vraisemblance", *Astronomy and Astrophysics*, **17**, 220–225, 1972.

19. M.O. Mennessier, "On statistical estimates from proper motions. III.", *Astronomy and Astrophysics*, **11**, 111–122, 1972.

20. G. Mersch and A. Heck, "Prediction of spectral classification from photometric observations — application to the $uvby\beta$ photometry and the MK spectral classification. II. General case", *Astronomy and Astrophysics*, **85**, 93–100, 1980.

21. J.F. Nicoll and I.E. Segal, "Correction of a criticism of the phenomenological quadratic redshift–distance law", *The Astrophysical Journal*, **258**, 457–466, 1982.

22. J.F. Nicoll and I.E. Segal, "Null influence of possible local extragalactic perturbations on tests of redshift–distance laws", *Astronomy and Astrophysics*, **115**, 398–403, 1982.

23. D.M. Peterson, "Methods in data reduction. I. Another look at least squares", *Publications of the Astronomical Society of the Pacific*, **91**, 546–552, 1979.

24. I.E. Segal, "Distance and model dependence of observational galaxy cluster concepts", *Astronomy and Astrophysics*, **123**, 151–158, 1983.

25. I.E. Segal and J.F. Nicoll, "Uniformity of quasars in the chronometric cosmology", *Astronomy and Astrophysics*, **144**, L23–L26, 1985.

5.6.2 Regression in General

1. P.R. Bevington, *Data Reduction and Error Analysis for the Physical Sciences*, McGraw–Hill, New York, 1969.

    (A recommendable text for regression and fitting, with many examples.)

2. N.R. Draper and H. Smith, *Applied Regression Analysis*, Wiley, New York, 1981 (2nd ed.).

3. B.S. Everitt and G. Dunn, *Advanced Methods of Data Exploration and Modelling*, Heinemann Educational Books, London, 1983.

    (A discursive overview of topics such as linear models and analysis of variance; PCA and clustering are also covered.)

4. M. Lybanon, "A better least–squares method when both variables have uncertainties", *American Journal of Physics* **52**, 22–26, 1984.

5. D.C. Montgomery and E.A. Peek, *Introduction to Linear Regression Analysis*, Wiley, New York, 1982.

6. G.A.F. Seber, *Linear Regression Analysis*, Wiley, New York, 1977.

7. G.B. Wetherill, *Elementary Statistical Methods*, Chapman and Hall, London, 1967.

    (An elementary introduction, with many examples.)

8. D. York, "Least squares fitting of a straight line", *Canadian Journal of Physics* **44**, 1079–1086, 1966.

    (Deals with the case of stochastic independent and dependent variables.)

5.6.3 Other Techniques

Regression excepted, little to date has been accomplished in astronomy using the other, varied, techniques discussed in this Chapter. In this, there clearly is scope for change! Some general references follow.

1. J.P. Benzécri, *L'Analyse des Données. II. L'Analyse des Correspondances*, Dunod, Paris, 1979 (3rd ed.).

    (The classical tome on Correspondence Analysis.)

2. W.W. Cooley and P.R. Lohnes, *Multivariate Data Analysis*, Wiley, New York, 1971.

   (See for Canonical Correlation Analysis, together with other techniques.)

3. A.D. Gordon, *Classification*, Chapman and Hall, London, 1981.

   (For some reading on multidimensional scaling, together with other topics.)

4. M. Greenacre, *Theory and Applications of Correspondence Analysis*, Academic Press, New York, 1984.

   (As the title suggests, a detailed study of Correspondence Analysis.)

5. J.B. Kruskal and M. Wish, *Multidimensional Scaling*, Sage, Beverly Hills, 1978.

6. L. Lebart, A. Morineau and K.M. Warwick, *Multivariate Statistical Analysis*, Wiley, New York, 1984.

   (This is especially recommendable for Multiple Correspondence Analysis.)

7. E. Malinvaud and J.C. Deville, "Data analysis in official socio–economic statistics", *Journal of the Royal Statistical Society Series A* **146**, 335–361, 1983.

   (The use of multivariate statistics on large data collections is discussed.)

8. S.S. Schifman, M.L. Reynolds and F.L. Young, *Introduction to Multidimensional Scaling*, Academic Press, New York, 1981.

9. W.S. Torgerson, *Theory and Methods of Scaling*, Wiley, New York, 1958.

   (Although old, this book is very readable on the subject of non–metric multidimensional scaling.)

10. M. Volle, *Analyse des Données*, Economica, Paris, 1981.

    (Correspondence Analysis is dealt with in a practical fashion.)

Chapter 6

Case Study: IUE Low Dispersion Spectra

6.1 Presentation

The aim of this case study is to illustrate the use of various methods introduced in previous chapters on a specific, real–life example. The emphasis here will be more on the statistical aspects of the problem than on the astrophysical details, which the interested reader can find in the specialized papers referenced and more particularly in Heck *et al.* (1984b, 1986a).

6.2 The IUE Satellite and its Data

The International Ultraviolet Explorer (IUE) satellite was launched on 26 January 1978 to collect spectra in the ultraviolet wavelength range (UV) for all types of celestial objects. To date, IUE can be qualified as the most successful and the most productive astronomical satellite. It is also the first "space telescope" to be exploited in the same way as a "mission" ground observatory, with visiting astronomers participating in real time in the decision loop for programming the observational sequences and collecting data during the observing shifts allocated to their approved programmes. At the time of writing, the satellite is still active 24 hours per day in geosynchronous orbit at about 40,000 km from the Earth. Its construction, launch and utilisation resulted from a joint venture by three space agencies: NASA (USA), SERC (UK) and ESA (Europe). It is exploited from two ground stations: one for 16 hours per day at NASA Goddard Space Flight Center (Greenbelt, Maryland, USA), and the other for the remaining 8 hours per day at the ESA VILSPA Satellite Tracking Station (Villafranca del Castillo, near Madrid, Spain).

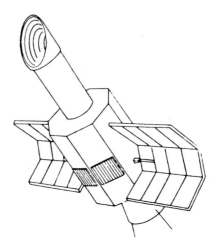

Figure 6.1: The International Ultraviolet Explorer (IUE).

IUE scientific instrumentation essentially consists of a Ritchey–Chrétien 45 cm telescope with two spectrographs: one working in the 1150 – 2000 Å range, and the other in the 1900 – 3200 Å range. Each spectrograph can record spectra in a low–resolution mode (≈ 7 Å) and in a high–resolution mode (≈ 0.1 Å). For more details on IUE, see Boggess *et al.* (1978a, 1978b), and see also the IUE Memorial Book (Kondo *et al.*, 1986) containing among others a chapter on UV spectral classification (Heck, 1986).

Only low–dispersion spectra will be considered in the following, and by spectrum we shall understand 410 flux values at 5 Å steps (just below the actual resolution) covering the whole IUE wavelength range (1150 – 3200 Å), obtained by merging the outputs of both spectrographs.

6.3 The Astrophysical Context

From earlier work on data collected by the S2/68 experiment on board the TD1 satellite (called "TD1" in the following), it had been shown that stars which were classified as spectrally normal in the visible range ($\approx 3500 - 4800$ Å) did not necessarily behave normally in the ultraviolet range and vice versa (see Cucchiaro *et al.*, 1978 and the references quoted therein). Consequently, MK spectral classifications which are defined in the visible range cannot simply be extrapolated to the UV.

IUE covers a larger wavelength range than TD1 (1150 – 3200 Å as against 1250 – 2550 Å) and it has, even at low dispersion, a resolution about five times

6.3. THE ASTROPHYSICAL CONTEXT

better than TD1 (7 Å as against 36 Å). Moreover IUE has observed a much broader range of stellar types than TD1 and has also reached significantly fainter magnitudes. Therefore a UV stellar classification programme was initiated in order to define, from IUE low-dispersion spectra, smooth UV spectral sequences. The latter were to describe stellar behaviour in the UV while staying as far as possible in accordance with the MK scheme in the visible.

The first volume of an *IUE Low-Dispersion Spectra Reference Atlas* (called the *Atlas* in the following) has already been produced (Heck et al., 1984a), together with reference sequences and standard stars, and a second volume devoted to peculiar groups is in preparation (see Heck et al., 1986b). The considerable classification work undertaken follows a classical morphological approach (Jaschek and Jaschek, 1984) and it essentially confirms that there is no one-to-one correspondence between the UV and visible ranges.

Let us recall here that the IUE spectral classification consists of a symbol expressing membership in a luminosity class (e.g. $s+$ defining supergiant, bright), followed by spectral-type symbols linked to the effective temperature of the star (e.g. $B5$).

Stellar spectral classifications are more than just taxonomical exercises aimed at labelling and categorizing stars by comparison with standards. They are used for describing fundamental physical parameters in the outer atmospheres of the stars, for discriminating peculiar objects, and for other subsidiary applications like distance determinations, interstellar extinction and population synthesis studies.

It is important to bear in mind that the classification systems are built independently of stellar physics in the sense that they are defined completely from spectral features, selected in standard stars, in a given wavelength range (see, e.g., Jaschek, 1979 and Morgan, 1984). If the schemes are based on a sufficiently large number of objects, then these classification schemes cannot be other than intimately linked with stellar physics. Such classification schemes will not necessarily relate to the same stellar layers, if they refer to different wavelength ranges. Consequently, the discrepancies found between the MK system and the UV framework are not surprising.

This also implies that the only way to confirm the correctness of the UV classification framework introduced in the *Atlas* is to remain in the same wavelength range. A statistical approach would moreover be independent of any *a priori* bias arising from existing schemes, either in the visible or in the ultraviolet ranges. An additional advantage of statistical methodology lies in the fact that it is able to work at will in a multidimensional parameter space, while classical morphological classifications rarely go beyond two dimensions.

The aim of this study is thus to apply multidimensional statistical algorithms to variables expressing as objectively as possible the information contained in the continuum and the spectral features of low-dispersion IUE stellar spectra. Several

	O	B	A	F	G	K
s+	3	16				
s	4	4	9	5	2	
s−	2	9				
g+	4	6	2			
g	9	23	4	5		
d+	5	13	3			
d	20	60	31	17	5	2

Table 6.1: UV spectral distribution of the stars sampled.

objectives may be pursued, but the most important will be the arrangement of the corresponding stars into groups, whose homogeneity, in terms of the classification symbolism introduced in the *Atlas*, should reflect the accuracy of this IUE UV classification scheme.

6.4 Selection of the Sample

Although much of the following is applicable to all types of spectra, consideration is restricted here to stars that are normal in the UV. Thus, for this application, we retained 264 IUE low–dispersion spectra which were technically good in the sense that images of poor quality (i.e. strongly underexposed or affected by saturation, microphonic noise or other defects) were discarded.

TD1 data had the advantage of resulting from a survey and of representing a magnitude–limited unbiased sample of bright stars, whereas IUE is pointed only at preselected targets from accepted proposals. It thereby provides a biased sample, mainly in favour of early (or hotter) spectral types (typical UV emitters). This is illustrated in Table 6.1, which gives the distribution of the sample stars for their UV spectral types (effective temperature decreasing from hotter O to cooler K stars) and luminosity classes (luminosity decreasing from bright supergiants, $s+$, to dwarfs, d).

Since we intend to compare flux values of stars with (sometimes very) different apparent magnitudes, it is necessary to normalize the flux scales. This was done by dividing each spectrum by the average flux value over the whole interval, which was equivalent to normalizing the integrated flux over the whole IUE range.

6.5 Definition of the Variables

Some of the algorithms will be applied below to the original 264×410 original flux values, but the 410 flux values will also be condensed into a smaller number of

6.5. DEFINITION OF THE VARIABLES

variables expressing as exhaustively as possible the information contained in the spectra. Clearly, the smaller the number of variables, the less computation time is required in the case of some of the multivariate data analysis algorithms.

Essentially two types of information can be extracted from a spectrum: on the one hand, the general shape of the continuum which will be described by an asymmetry coefficient; and on the other hand, the various data relative to the individual spectral lines. In total, 60 line intensity values were used.

6.5.1 The Continuum Asymmetry Coefficient

To define the continuum asymmetry coefficient, the spectra were first smoothed by binning the fluxes into 31 boxes of 65 Å. These correspond to 403 intervals of the initial 5 Å steps, and cover the interval 1170 – 3185 Å (the fractions of spectra dropped at each end are insignificant). Such figures seemed to be an excellent compromise between too many bins (with too much influence from the spectral lines) and too few (with consequent alteration of the general continuum shape). In each bin, the flux was represented by the median which turned out to be more suitable than the mean, the latter being too much affected by possible lines in the bin.

The following asymmetry coefficient was then calculated:

$$S = (A_1 - A_2)/(A_1 + A_2)$$

where A_1 and A_2 are the areas illustrated in Figure 6.2 and correspond respectively to the ranges 1430 – 1885 Å and 2730 – 3185 Å. The first interval was selected as the largest feasible interval between possible parasitic effects arising in the spectra from:

- on the shortwavelength side, either a $L\alpha$ geocoronal emission or strong stellar absorption;

- on the longwavelength side, a 2200 Å interstellar absorption bump.

There were seven 65 Å bins between these two features. Area A_2 was taken with the same length at the longwavelength end of the spectrum, and as far as possible from the 2200 Å bump. The positions and sizes of A_1 and A_2 could be defined differently, but the choices made here were fully satisfactory for our purposes.

It is easy to see that S varies from 1 (the extreme case of hot stars — all the flux at the shortwavelength end) to -1 (the extreme case of cool stars — all the flux at the longwavelength end). Both flat and balanced spectra give $S = 0$.

The asymmetry coefficient appeared immediately to be a powerful discriminating parameter for normal stars. However, the reddening turned out to be more troublesome than expected and it had to be taken into account, even outside the

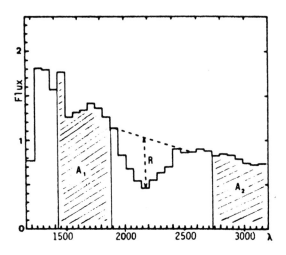

R: the reddening depth.
A1 and A2: areas.

Figure 6.2: Illustration of terms in the asymmetry coefficient S (see text).

zone that we avoided. Actually, a simple glance at Seaton's (1979) interstellar–extinction curve (his Figure 1) makes it clear that the absorption is as strong around 1350 Å as around 2200 Å. This means that a hot–star spectrum (clearly asymmetric with S close to 1 when unreddened) becomes fairly symmetric (with S close to 0) when reddened (see Figure 6.3).

6.5.2 The Reddening Effect

In order to correct the asymmetry coefficient for the reddening effect, we introduced a new parameter R (see Figure 6.2), which we called the reddening depth and defined as:

$$R = (F_{1890} + F_{2530})/2 - F_{2210}$$

(where F_λ represents the flux at wavelength λ) giving a rough measurement of the amplitude of the 2200 Å bump.

We then dereddened a number of spectra from our sample (chosen at various degrees of reddening) by using Seaton's (1979) absorption law and compared the pairs (reddened–dereddened) in the normalized representation. As demonstrated by Figure 6.3, the reddening increases area A_2 compared to area A_1, which stays roughly at the same level in most cases.

A rather simple functional relationship could be empirically established be-

6.6. SPECTRAL FEATURES

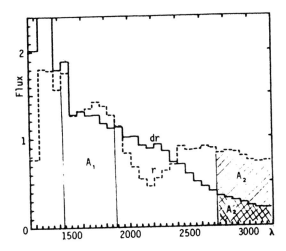

```
r    denotes the reddened curve,
dr   denotes the de-reddened curve.
A2   is actually smaller in the dereddened spectrum.
```

Figure 6.3: Illustration of the reddening effect on a normalized spectrum.

tween the reddening depth R and the difference in area A_2 due to the reddening:

$$\ln(A_2 - A'_2) = 3.0 \ln R - 0.5$$

where A_2 corresponds to the reddened spectrum and A'_2 to the unreddened one. For $R = 0$, we have of course $A_2 = A'_2$. Quite naturally, the coefficients of this relationship could be refined with a bigger sample.

Thus, when a positive reddening is detected, area A_2 has to be decreased by $(A_2 - A'_2)$ before calculating the asymmetry coefficient S.

It should be emphasized here that, although only one reddening law has been used for illustrating and for determining a rough empirical relationship, this does not imply that we consider it of universal application in the UV. It is however sufficient within the overall accuracy of this case study, whose results are mainly qualitative.

6.6 Spectral Features

6.6.1 Generalities

We decided to concentrate on spectral features which are classically used for spectral classification, i.e. emission and/or absorption lines which can play a rôle with

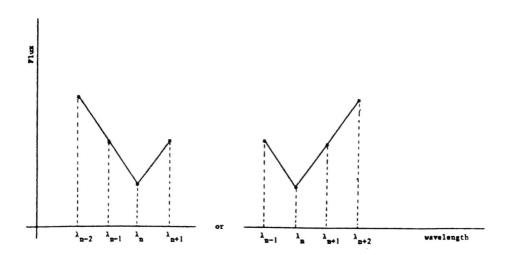

Figure 6.4: Test used for objectively detecting potential lines in the spectra.

different degrees of sophistication: firstly, by their presence or absence only; secondly, by their comparative intensity; and finally, by their intensity weighted in an appropriate way.

In any case, an objective approach should begin by a systematic search for these lines. As our sample consists only of normal stars, we consider in the following only absorption lines but, *mutatis mutandis*, similar considerations could also apply to emission features.

6.6.2 Objective Detection of the Spectral Lines

We looked for the most frequent minima in the spectra of our sample. Various morphological tests were applied to the values of each original flux vector by moving sets of 3 to 7 points. Then a histogram was constructed to indicate the frequency of the minima detected over the whole spectral range and over the whole sample. The test which appeared to be the most adequate and the most efficient (in terms of the best compromise between line detection and insensitivity to noise) is reproduced in Figure 6.4. In other terms, we considered as a potential line wavelength, that corresponding to the flux component n such that:

$$F_{n-2} > F_{n-1} > F_n < F_{n+1}$$

or such that

$$F_{n-1} > F_n < F_{n+1} < F_{n+2}$$

6.6. SPECTRAL FEATURES

1175*	1565	2065	2665*
1215*	1610*	2100	2695
1230	1625*	2145	2720
1260*	1640*	2300	2750*
1275	1655	2345	2800*
1300*	1670	2365	2835
1330*	1720*	2390	2855*
1365*	1765	2410	2875
1395*	1810	2440	2930
1415	1850*	2475	2990
1430*	1895*	2490	3025
1455*	1925*	2525	3065
1470*	1960*	2540	3105
1535	1995	2610	3120
1550*	2040	2635	3160

Asterisks indicate lines used for classification
in the IUE Normal Stars Atlas (Heck et al., 1984a).

Table 6.2: Bin wavelengths corresponding to the 60 most frequent lines in the spectral sample at hand.

where F_i represents the flux at the ith component. Such an asymmetry in the tests permits the detection of lines in the wings of larger nearby ones.

We decided to select the most frequent lines because we were dealing with a sample of normal stars. With peculiar stars, the criterion would be different since one given line appearing only in a few stars could be a discriminator for that particular group. The approximate wavelength (5 Å precision) of the 60 most frequent lines are given in Table 6.2. They include all the lines used in the morphological classification given in the *Atlas*.

Additional lines could have been used, but 60 turned out to be a good figure for this study. The line selection was in principle done above the noise level corresponding to a purely random process. However, the minima corresponding to the IUE réseau marks were ignored, as were those resulting from, on the one hand, the junction of the "long-" and "shortwavelength" parts of the spectra and, on the other hand, from the camera sensitivity drops towards the extremities of their respective spectral ranges.

6.6.3 Line Intensities

Line intensities were calculated (from normalized flux values) as the differences between the minima and associated continua smoothed at the median values over 13 components centred on each line.

Allowing for the usual difficulties of drawing a spectral continuum, we believe this rough systematic procedure does not introduce additional imprecision in the whole approach and does not modify substantially the essentially qualitative results of the study.

6.6.4 Weighting Line Intensities

In order to enhance the importance of lines in the shortwavelength range for hot stars (where the signal is most significant), and similarly of lines in the longwavelength range for cool stars, the line intensities were weighted with the "variable Procrustean bed" (VPB) technique (introduced in Heck et al., 1984b) through the formula

$$D'_i = D_i(1 + S - 2S(L_i - 1155)/2045)$$

where D_i is the unweighted depth of the ith line, S is the asymmetry coefficient, and L_i is the wavelength (Å) of the ith line ($i = 1$ to 60).

As illustrated in Figure 6.5, this corresponds, for an extreme hot star (curve H), to multiplying the shortwavelength–side line intensities by 2 and the longwavelength–side ones by 0. Those in between have a weight progressively decreasing from 2 to 0 as the wavelength increases. The situation is reversed for an extreme cool star (curve C), and all intermediate cases are weighted intermediately (in the sense of the arrows) in a similar way. The neutral case (all line intensities weighted by 1) corresponds to $S = 0$.

Here again, in a statistical parody of the deplorable mania of the Greek mythological robber Procrustes, the line intensities were forced to comply with the asymmetry coefficient (Procrustes' bed) which is variable with the star at hand. In other words, while a statistical variable generally receives the same weight for all individuals of a sample, in this application the weight is a function of the individuals themselves.

Finally it may be noted that the number of variables used could be even more reduced by looking at correlations between those defined, and rejecting redundant variables from consideration.

6.7 Multivariate Analyses

We are given a 264×403 table, consisting of 264 spectra each measured on 403 flux values (slightly reduced from the original 410 flux values, by dropping insignificant

6.7. MULTIVARIATE ANALYSES

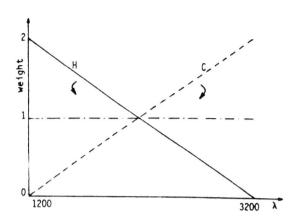

The abscissa represents increasing wavelength.
Curve H corresponds to the extreme case of hot stars (S = 1),
 and curve C to the extreme case of cool stars (S = -1).
The horizontal line corresponds to S = 0.

Figure 6.5: Weighting of the line intensities by the asymmetry coefficient through the Variable Procrustean Bed technique.

extremities of the spectra: cf. section 6.5.1 above). We also have a 264 × 61 table, the parameters being the asymmetry coefficient and 60 weighted line intensity values. As a general rule, as analyses are carried out, it may be interesting or even necessary to discard anomalous spectra or confusing parameters. It may similarly be of interest to redo the initial coding of the data, to make use of data transformations, and so on.

6.7.1 Principal Components Analysis

A Principal Components Analysis (PCA) condenses the dimensionality of a parameter space. It may be used as a data "cleaning" stage, prior to a clustering. In the case of the 264 × 403 table of flux values, the 264 spectra were *centred*: i.e. the PCA was carried out on a 403 × 403 covariance matrix. The percentage variance explained by the first axis was 77.5%, by the first plane 89%, and by the best fitting 3–dimensional subspace 94%. The marked linearity of the data was identified as being temperature–related, but also had some luminosity effects. Greater negativity on the first axis was essentially associated with cooler stars and, for a given temperature, more luminous stars.

The PCA of the 264 × 61 table offers computational advantages, and because of the carefully selected parameters may be of greater interpretative value. The spectra were again centred (i.e. leading to a PCA of the 61 ×61 covariance matrix; an alternative of a PCA of the correlation matrix was not undertaken since it would destroy the effects of the VPB weighting procedure — cf. section 6.6.4 above). The following percentages of variance explained by the seven successive principal components were obtained: 49, 18, 9, 5, 3.5, 2.5, and 1.5. Hence the best fitting seven–dimensional subspace accounts for 88% of the variance of the 264 spectra in the 61 parameter space.

Regarding interpretation of the principal components, the first axis was strongly correlated with the $L\alpha$ line (stellar absorption), followed by the asymmetry coefficient, and by other lines that were all indicators of spectral type. Thus this first axis was essentially an effective temperature discriminator.

The second axis was most correlated with a line that had not been retained for the *Atlas* morphological classification because it was located close to other important lines. The three following ones discriminate mainly the effective temperature. The fifth line was not selected for the *Atlas* either.

The third and fourth axes were essentially discriminators of effective temperature among the cool types, while the fifth axis discriminated luminosity among hot stars.

6.7.2 Cluster Analysis

Among the cluster analyses carried out was a clustering on the seven principal components resulting from the PCA of the 264×61 matrix. Hierarchical clustering was used because, not knowing the inherent number of clusters, a range of groups could be studied (the 30, 40, and 50 group solutions, in particular, were examined). The minimum variance criterion was used for the desired objective of determining synoptic groups.

The results are shown in Table 6.3. The first column is the star identifier, the second is the IUE spectral classification (this symbolism is used in the *Atlas*) and the third, fourth and fifth columns are respectively the 30, 40 and 50 group solutions.

HD 8890	s F8	1	1	1
HD 79447	d+B3	2	2	2
HD 36824	d B2.5	2	2	2
HD 192685	d B2.5	2	2	2
HD 32630	d B3	2	2	2

6.7. MULTIVARIATE ANALYSES 185

HD 37674	d B3	2	2	2
HD 168905	d B2.5	2	2	2
HD 120315	d B3	2	2	2
HD 64802	d B2.5	2	2	2
HD 4142	d B4	2	2	2
HD 829	d+B3	2	2	2
HD 100600	d B2.5	2	2	2
HD 42690	d B2	2	2	2
HD 36629	d B2	2	2	2
HD 190993	d B2.5	2	2	2
HD 37776	d B2	2	2	2
HD 61831	d B2.5	2	2	2
HD 37129	d+B2.5	2	2	2
HD 86440	s-B5	3	3	3
HD 53138	s+B3	3	3	3
HD 2905	s+B1	3	3	3
HD 164353	s-B5	3	3	3
HD 23408	g B6	3	3	3
HD 83183	g+B6	3	3	3
HD 51309	g+B3	3	3	3
HD 46769	d B6	3	3	3
CPD $-72°1184$	g B0	3	3	49
HD 34078	d+O9.5	3	3	49
HD 152233	g O5	4	4	4
HD 93403	g O5	4	4	4
HD 210839	s O5	4	4	4
HD 152248	s O7	4	4	4
HD 46223	d O4	4	4	4
HD 46150	d O5	4	4	4
HD 47129	g+O7	4	4	4
HD 162978	g+O7	4	4	45
HD 48099	d O7	4	4	45
HD 165052	d O7	4	4	45
HD 213558	d A1	5	5	5
HD 87737	s A0	5	5	5
HD 41695	d A1	5	5	5
HD 156208	g A1	5	5	5
HD 86986	d A1	5	5	5
HD 29646	d A1	5	5	5
HD 20346	g:A2	5	5	5
HD 80081	d A2	5	5	5
HD 166205	d A2	5	5	5
HD 104035	s A2	5	5	5
HD 177724	d A0	5	5	5
HD 48250	d A2	5	5	5
HD 111775	d+A0	5	5	5
HD 58142	d+A1	5	5	5
HD 60778	d A2	5	5	5
HD 149212	g B9	5	5	5

HD 137422	g+A2	5	5	5
HD 62832	d A1	5	5	5
HD 11031	d A2	5	5	5
HD 167838	s+B5	5	5	5
HD 103287	d A0	5	5	5
HD 95418	d A1	5	5	5
HD 27962	d A2	5	5	5
HD 9132	d A1	5	5	5
HD 12301	s-B8	5	5	5
HD 166937	s+B8	5	5	5
HD 53367	g B0	5	5	43
HD 198478	s+B3	5	5	43
HD 77581	s+B0.5	5	5	43
HD 190603	s B2	5	5	43
HD 41117	s+B2	5	5	43
HD 199216	s-B2	5	5	43
HD 148379	s+B2	5	5	43
HD 206165	s-B2	5	5	43
HD 163181	s B1	5	5	50
HD 29335	d B6	6	6	6
HD 147394	d B5	6	6	6
HD 210424	d B7	6	6	6
HD 23480	d B6	6	6	6
HD 25340	d B5	6	6	6
HD 22928	g B5	6	6	6
HD 197392	g B8	6	6	6
HD 21071	d B7	6	6	6
HD 90994	d B6	6	6	6
HD 162374	d B6	6	6	6
HD 199081	(example)	6	6	6
HD 31512	d B6	6	6	6
HD 37903	d B1.5	6	6	41
HD 37367	d+B2.5	6	6	41
HD 27396	d B4	6	6	41
HD 23060	d B3	6	6	41
HD 4727	d+B4	6	6	41
HD 34759	d B4	6	6	41
HD 83754	d B5	6	6	41
HD 52942	d B1.5	6	6	41
HD 142983	g B4	6	6	47
HD 50846	g+B5	6	6	47
HD 47755	d B3	6	6	47
HD 60753	d B3	6	6	47
HD 40136	d F2	7	7	7
HD 128167	d F3	7	7	7
HD 20902	s F5	7	7	7
HD 173667	d F6	7	7	7
HD 61421	d F5	7	7	7
HD 77370	d F4	7	7	7

6.7. MULTIVARIATE ANALYSES 187

HD 113139	d F2	7	7	7
HD 99028	g F3	7	7	7
HD 152667	s+B0.5	8	8	8
HD 47240	s+B1	8	8	8
HD 24398	s-B1	8	8	8
BD -9°4395	s B1.5	8	8	8
HD 47432	s+O9	8	36	36
HD 122879	s+B0	8	36	36
HD 37043	d+O9	9	9	9
HD 47839	d O8	9	9	9
HD 13268	g O8	10	10	10
HD 188209	s-O9.5	10	10	10
HD 30614	s+O9	10	10	10
BD +60°497	d O7	10	10	10
HD 48329	s G8	11	11	11
HD 16901	s G0	11	39	39
HD 93028	d+O9.5	12	12	12
HD 40111	g B1	12	12	12
HD 14633	d O9	12	12	12
HD 37061	d+B0.5	12	12	12
HD 218376	g B1	12	12	12
HD 3360	d+B1.5	13	13	13
HD 200120	d B0.5	13	13	13
HD 37023	d B0	13	13	13
HD 52918	d+B1	13	13	13
HD 63922	d+B0	13	13	13
HD 75821	g B0	13	13	13
HD 37744	d B1.5	13	13	44
HD 37020	d B0	13	13	44
HD 37042	d B0.5	13	13	44
HD 31726	d B2	13	13	44
HD 183143	s+B7	14	14	14
HD 147084	g+A5	14	14	14
HD 21291	s B9	14	14	14
HD 5448	d A4	14	14	14
HD 122408	d:A3	14	14	14
HD 122408	d:A3	14	14	14
HD 24432	s A0	14	14	14
HD 79439	d A5	14	14	14
HD 199478	s+B8	14	14	14
HD 17138	d A3	14	14	14
HD 21389	s A0	14	14	14
HD 6619	d A5	14	14	42
HD 40932	d A4	14	14	42
HD 116842	d A5	14	14	42
HD 197345	s A2	14	14	42
HD 76644	d A7	14	14	42
HD 216701	d A5	14	14	42
HD 189849	g A5	14	14	42

HD 87696	d A7	14	14	42
HD 11636	d A5	14	14	42
HD 216956	d A3	14	14	42
HD 76576	d:A5	14	14	42
HD 74180	s F2	14	14	42
HD 159561	g A5	14	14	42
HD 22049	d K	14	14	48
HD 193682	g:O4	15	15	15
HD 215835	d O5	15	15	15
HD 192281	d O4	15	15	15
HD 168076	d O4	15	15	15
HD 15570	s O4	15	15	15
HD 15629	d O5	15	15	15
HD 170153	d F6	16	16	16
HD 10700	d G	16	35	35
HD 85504	d+A0	17	17	17
HD 38206	d A0	17	17	17
HD 23850	g B8	17	17	17
HD 29365	d B8	17	17	17
HD 61429	d+B7	17	17	17
HD 108767	d B9	17	17	17
HD 58350	s+B5	17	17	17
HD 23753	g B8	17	17	17
HD 107832	d+B8	17	17	17
HD 10250	d B9	17	17	17
HD 23630	g B7	17	17	17
HD 148605	d B2	17	17	17
HD 201908	d B8	17	17	17
HD 23432	d B8	17	17	17
HD 86360	d+B9	17	17	17
HD 223778	d K3	18	18	18
HD 36512	d B0	19	19	19
HD 214680	d O9	19	19	19
HD 34816	d B0.5	19	19	19
HD 55857	d B0.5	19	19	19
HD 38666	d O9	19	19	19
HD 57682	d O9.5	19	19	19
HD 74273	d B1.5	19	34	34
HD 144470	d B1	19	34	34
HD 212571	d B1	19	34	34
HD 46056	d O8	20	20	20
HD 46149	d+O8	20	20	20
HD 52266	g O9	20	20	20
HD 46202	d+O9	20	20	20
HD 53974	g B0.5	20	20	20
HD 209481	g O9	20	20	20
HD 58946	d F2	21	21	21
HD 59612	s A5	21	21	21
HD 36673	s F0	21	21	21

6.7. MULTIVARIATE ANALYSES 189

HD 182640	d:F2	21	21	21
HD 27290	d F1	21	21	21
HD 90589	d F3	21	21	21
HD 65456	d A5	21	21	21
HD 27176	d F0	21	21	21
HD 147547	g F0	21	21	21
HD 161471	s F3	21	21	21
HD 12311	g F0	21	21	21
HD 206901	d F4	21	31	31
HD 89025	g F2	21	31	31
HD 202444	d:F3	21	31	31
HD 78362	g F3	21	31	31
HD 127739	d F3	21	31	31
HD 210221	s A3	21	33	33
HD 90772	s A5	21	33	33
HD 148743	s A7	21	33	33
HD 128620	d:G	22	22	22
HD 6582	d G	22	32	32
HDE 326330	g B1:	23	23	23
HD 64760	s-B0.5	23	23	23
HD 91316	s-B1	23	23	23
HD 123008	s+O9	23	23	23
HD 152249	s-O9	23	37	37
HD 152247	g+O9	23	37	37
HD 54439	d:B1.5	24	24	24
HD 52721	d B2:	24	24	24
HD 200310	d B1	24	24	24
HD 154445	d B1	24	24	24
HD 54306	d B1	24	24	24
HD 147933	d B1.5	24	24	24
HD 207330	g B2.5	24	38	38
HD 92741	g+B1	24	38	38
HD 149881	g:B0.5	24	38	38
HD 150898	g+B0:	24	38	38
HD 165024	g+B1	24	38	38
HD 51283	g B2	24	38	38
HD 219188	g B0.5	24	38	38
HD 173502	g B1	24	38	38
HD 167756	s-B0.5	24	38	38
HD 164794	d O3	25	25	25
HD 93205	d O3	25	25	25
HD 93204	d:O3	25	25	25
HD 93250	d O3	25	25	46
CPD-59°2600	d:O	25	25	46
HDE 303308	d O3	25	25	46
HD 93130	g O5	25	25	46
BD +60°594	g:O8	26	26	26
BD +63°1964	g B0	26	26	26
HD 40893	g O9.5	26	26	26

CPD −41° 7711	d B1	26	40	40
HD 164402	s−B0	27	27	27
HD 57061	g+O9.5	27	27	27
HD 167264	s+B0	27	27	27
HD 10307	d G1.5	28	28	28
HD 102870	d F9	28	28	28
HD 20630	d G5	28	28	28
BD +60° 2522	s O7	29	29	29
HD 109387	d+B7	30	30	30
HD 23302	g B6	30	30	30
HD 87901	d B7	30	30	30
HD 183914	d B8	30	30	30
HD 47054	d B8	30	30	30
HD 23324	d B8	30	30	30

Table 6.3: Star identifiers, IUE spectral classification, and 30-, 40- and 50-cluster solutions.

Note that the group sequence numbers used in the final three columns of Table 6.3 have no inherent meaning. Sorting of star identifiers was carried out to make the grouping clear, but in other respects has no inherent significance. The hierarchical structure between the 30, 40 and 50 cluster solutions (produced by the one algorithm) can be seen. It is inherent in the clustering algorithm used that the 50-group solution (the rightmost column of Table 6.3) is both a finer grouping and gives a more homogenous set of clusters than the 40 and 30 group solutions. The number of clusters looked at (i.e. 30, 40 and 50) were selected as reasonable figures, given the number of stars per group and the spectral classification (shown in Column 2 of Table 6.3, and now to be used for comparison).

Relative to spectral classification (Column 2), as the number of groups increases from 30 to 50, bigger groups are split into smaller more homogenous groups. This homogeneity is limited, firstly, by the relatively small size of the sample at hand (i.e. 264 stars) compared to the number of possible spectral classifications; and, secondly, by the underlying continuous physical variables (viz. effective temperature and intrinsic luminosity).

The overall approach developed here could also be used to predict UV spectral classification by assimilation. Stars with unknown UV spectral classification could be introduced into the general sample and the procedure applied to the whole population. The predicted classification could be taken as the spectral *mode* of the group to which the star has been assigned by the algorithm. The precision obtained would then be of the order of the *ranges* in spectral type and luminosity class of these groups.

As an example, star HD 199081 (on page 186) is in group number 6. The range here is $d - g$ (dwarf – giant) in luminosity, and B5–B7 in spectral type. The mode is *d B6*. Hence in the absence of further information, we would assign this star to spectral class *d B6*, with a tolerance indicated by the above-mentioned interval.

6.7.3 Multiple Discriminant Analysis

Cluster Analysis allowed good discrimination between spectral types, but was somewhat less effective in discriminating between luminosity classes alone. Hence, Multiple Discriminant Analysis was used to assess the discrimination between these classes. The following luminosity classes were used: *s* (supergiant), *g* (giant) and *d* (dwarf) (the star discussed at the end of section 6.7.2, Cluster Analysis, being assigned to *d*).

It was found that the three discriminant factors obtained allowed pairwise discrimination between these classes. Figure 6.6 shows a plot of *d* versus *s* in the plane defined by discriminant factors 1 and 2; while Figure 6.7 shows the plot obtained of discriminant factors 1 and 3 where the discrimination between *g* and *s* can be seen. Unclear cases lie close to separating lines, and unknown cases may be decided upon by investigating the relative position in the discriminant factor space.

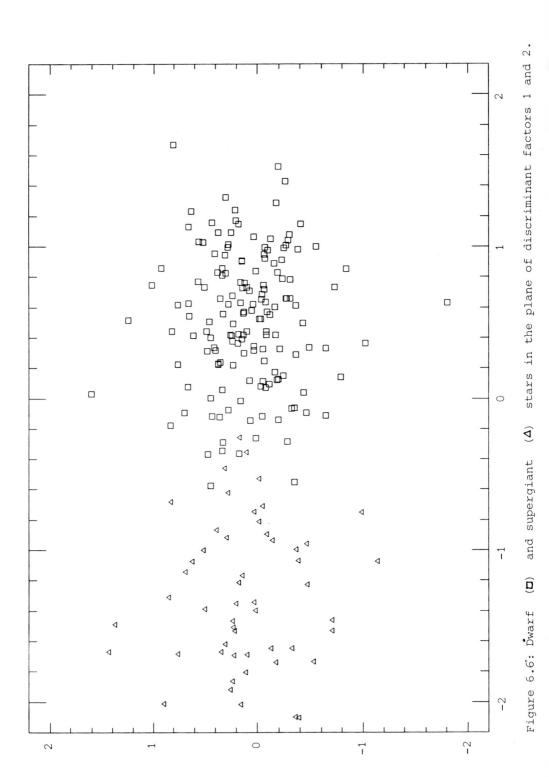

Figure 6.6: Dwarf (☐) and supergiant (△) stars in the plane of discriminant factors 1 and 2.

6.7. MULTIVARIATE ANALYSES

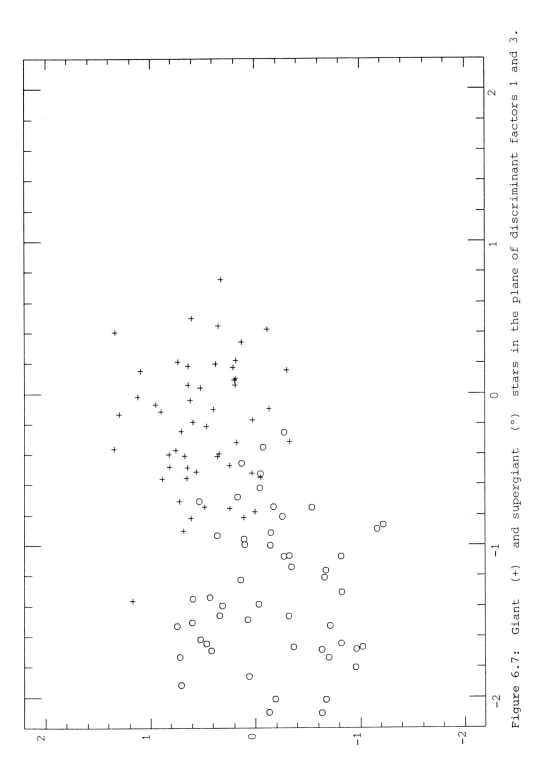

Figure 6.7: Giant (+) and supergiant (°) stars in the plane of discriminant factors 1 and 3.

Chapter 7

Conclusion: Strategies for Analysing Data

In this concluding chapter, we will summarize the options involved in applying methods studied in earlier chapters. Whether a method will perform well in a practical situation will depend both on the geometrical/mathematical nature of the method employed, and simultaneously on the quality and type of the data analysed. We will review the former first.

7.1 Objectives of Multivariate Methods

1. Principal Components Analysis:

   - Reducing the dimensionality of the parameter space to its inherent dimensionality.
   - Thereby also eliminating "noise", "cleaning" the data, and lessening the volume of data.
   - Determining the most important linear combinations of the parameters (variables).
   - Using these to determine linear combinations, and even quadratic, logarithmic, and other, combinations.
   - Determining the important parameters present.
   - Determining "latent", underlying variables present.
   - Visualizing the data by the selection of the most important planar views of it.
   - Identification of groups of objects or of variables.
   - Identification of outliers or anomalous objects/variables.

2. Cluster Analysis:
   - Determining groups of objects, using some criterion.
   - Without prior knowledge of what groups are present, obtaining a range of groupings and using this to assess the true group structure.

3. Discriminant Analysis:
   - Assessing the separability of known groups.
   - Assigning objects to one of a number of known groups.
   - Displaying the groups, optimally separating the known classes.

4. Correspondence Analysis:
   - Identifying simultaneously groups of objects and variables.
   - Determining "latent", underlying variables.
   - Determining the important variables present.
   - Identifying outliers or anomalous objects/variables.

5. Principal Coordinates Analysis:
   - Determining groups, and displaying.

6. Canonical Correlation Analysis:
   - Assessing relationships of two groups of variables relating to the same object population.

7. Regression Analysis:
   - Fitting of a straight line to a set of points and assessing the significance of anomalous points.

7.2 Types of Input Data

1. Principal Components Analysis implicitly uses Euclidean distances: real valued data should therefore be used.

2. Clustering uses either distances or dissimilarities. Either may be calculated from given data. Clustering routines may accept the latter as input, or may work on distance/dissimilarity input.

3. Most of the Discriminant Analysis methods looked at assume real valued data.

7.3. STRATEGIES OF ANALYSIS

4. Correspondence Analysis allows a wide range of data types. It is suitable for data in the form of frequencies of occurence. It also is suitable for data in a form of coding which associates response categories with all possible data values (this latter is particularly interesting for diverse — qualitative/real and categorical/integer — types of data).

5. Principal Coordinates Analysis assumes distances as input. Other multidimensional scaling methods assume dissimilarities.

6. Canonical Correlation Analysis assumes real valued data.

7. Regression Analysis, as discussed in this book, assumes real valued data.

7.3 Strategies of Analysis

The complimentary use of multivariate methods often offer powerful ways of extracting information from the data. The following are some possibilities.

1. The use of clustering and of a display technique (Principal Components Analysis, Correspondence Analysis, etc.) can give complimentary views of the same data set, and can lead to more robust conclusions about the data. The latter, for example, look at groupings relative to projections onto axes, whilst the former looks at fully multidimensional grouping.

2. A Cluster Analysis can be carried out on the more informative first few axes of a Principal Components Analysis.

3. A Principal Components Analysis can be used on large clusters resulting from a Cluster Analysis in order to help interpret them.

4. Principal Components Analysis or Cluster Analysis may be used to define groups, and Discriminant Analysis may be used to subsequently assess them.

5. For complex data sets, a recoding can be carried out so that Correspondence Analysis can be used; or distances/dissimilarities can be defined so that clustering or multidimensional scaling can be employed.

6. Prior to analysis, the analyst should examine the correlation matrix in the case of Principal Components Analysis, or the Burt table in the case of Multiple Correspondence Analysis. All findings should be verified as far as possible with respect to the given, original data.

General References

The following have been referenced in the text. For bibliographies relating to the different techniques studied, see the relevant sections in the respective chapters.

1. H.-M. Adorf, "Classification of low-resolution stellar spectra via template matching — a simulation study", *Data Analysis and Astronomy*, (Proceedings of International Workshop on Data Analysis and Astronomy, Erice, Italy, April 1986) Plenum Press, New York (1986, forthcoming).

2. M.R. Anderberg, *Cluster Analysis for Applications*, Academic Press, New York, 1973.

3. P.R. Bevington, *Data Reduction and Error Analysis for the Physical Sciences*, McGraw-Hill, New York, 1969.

4. R.K. Blashfield and M.S. Aldenderfer, "The literature on cluster analysis", *Multivariate Behavioral Research*, **13**, 271–295, 1978.

5. A. Boggess, A. Carr, D.C. Evans, D. Fischel, H.R. Freeman, C.F. Fuechsel, D.A. Klinglesmith, V.L. Krueger, G.W. Longanecker, J.V. Moore, E.J. Pyle, F. Rebar, K.O. Sizemore, W. Sparks, A.B. Underhill, H.D. Vitagliano, D.K. West, F. Macchetto, B. Fitton, P.J. Barker, E. Dunford, P.M. Gondhalekar, J.E. Hall, V.A.W. Harrison, M.B. Oliver, M.C.W. Sandford, P.A. Vaughan, A.K. Ward, B.E. Anderson, A. Boksenberg, C.I. Coleman, M.A.J. Snijders and R. Wilson, "The IUE spacecraft and instrumentation", *Nature*, **275**, 372–377, 1978a.

6. A. Boggess, R.C. Bohlin, D.C. Evans, H.R. Freeman, T.R. Gull, S.R. Heap, D.A. Klinglesmith, G.R. Longanecker, W. Sparks, D.K. West, A.V. Holm, P.M. Perry, F.H. Schiffer III, B.E. Turnrose, C.C. Wu, A.L. Lane, J.L. Linsky, B.D. Savage, P. Benvenuti, A. Cassatella, J. Clavel, A. Heck, F. Macchetto, M.V. Penston, P.L. Selvelli, E. Dunford, P.M. Gondhalekar, M.B. Oliver, M.C.W. Sandford, D.J. Stickland, A. Boksenberg, C.I. Coleman, M.A.J. Snijders and R. Wilson, "In-flight performance of the IUE", *Nature*, **275**, 377–385, 1978b.

7. J.M. Chambers, *Computational Methods for Data Analysis*, Wiley, New York, 1977.

8. A. Cucchiaro, M. Jaschek and C. Jaschek, *An Atlas of Ultraviolet Stellar Spectra*, Liège and Strasbourg, 1978.

9. E. Davoust and W.D. Pence, "Detailed bibliography on the surface photometry of galaxies", *Astronomy and Astrophysics Supplement Series*, **49**, 631–661, 1982.

10. B.S. Everitt, *Cluster Analysis*, Heinemann Educational Books, London, 1980 (2nd ed.).

11. A.D. Gordon, *Classification*, Chapman and Hall, London, 1981.

12. J.C. Gower and P. Legendre, "Metric and Euclidean properties of dissimilarity coefficients" *Journal of Classification* **3**, 5–48, 1986.

13. P.A.V. Hall and G.R. Dowling, "Approximate string matching", *Computing Surveys*, **12**, 381–402, 1980.

14. D.J. Hand, *Discrimination and Classification*, Wiley, New York, 1981.

15. A. Heck, "UV stellar spectral classification", in Y. Kondo et al. (eds.), *Scientific Accomplishments of the IUE*, D. Reidel, Dordrecht, 1986 (in press).

16. A. Heck, D. Egret, M. Jaschek and C. Jaschek, *IUE Low Resolution Spectra Reference Atlas — Part 1. Normal Stars*, European Space Agency Special Publication 1052, 1984a.

17. A. Heck, D. Egret, Ph. Nobelis and J.C. Turlot, "Statistical classification of IUE stellar spectra by the Variable Procrustean Bed (VPB) approach", in *Fourth European IUE Conference*, European Space Agency Special Publication 218, pp. 257–261, 1984b.

18. A. Heck, D. Egret, Ph. Nobelis and J.C. Turlot, "Statistical confirmation of the UV stellar classification system based on IUE low–dispersion stellar spectra", *Astrophysics and Space Science* **120**, 223–237, 1986a.

19. A. Heck, D. Egret, B.J.M. Hassall, C. Jaschek, M. Jaschek and A. Talavera, "IUE low–dispersion spectra reference atlas — Volume II. Peculiar stars" in *New Insights in Astrophysics — Eight Years of UV Astronomy with IUE*, European Space Agency Special Publication 263, 661–664, 1986b.

20. C. Jaschek, "Classification spectrale" in *Classification Stellaire*, Ecole de Goutelas, ed. D. Ballereau, Observatoire de Meudon, 1979.

21. M. Jaschek and C. Jaschek, "Classification of ultraviolet spectra", in *The MK Process and Stellar Classification*, ed. R.F. Garrison, David Dunlap Observatory, 290, 1984.

22. Y. Kondo, A. Boggess, M. Grewing, C. de Jager, A.L. Lane, J.L. Linsky, W. Wamsteker and R. Wilson (eds.), *Scientific Accomplishments of the IUE*, D. Reidel, Dordrecht, 1986 (in press).

23. J.B. Kruskal, "An overview of sequence comparison: time warps, string edits, and macromolecules", *SIAM Review*, **25**, 201–237, 1983.

24. J.B. Kruskal and M. Wish, *Multidimensional Scaling*, Sage, Beverly Hills, 1978.

25. M.J. Kurtz, "Classification methods: an introductory survey", in *Statistical Methods in Astronomy*, European Space Agency Special Publication 201, 47–58, 1983.

26. O. Lefèvre, A. Bijaoui, G. Mathez, J.P. Picat and G. Lelièvre, "Electronographic BV photometry of three distant clusters of galaxies", *Astronomy and Astrophysics*, **154**, 92–99, 1986.

27. L. Lebart, A. Morineau and K.M. Warwick, *Multivariate Statistical Analysis*, Wiley, New York, 1984.

28. T.E. Lutz, "Estimation — comments on least squares and other topics", in *Statistical Methods in Astronomy*, European Space Agency Special Publication 201, 179–185, 1983.

29. M. Lybanon, "A better least-squares method when both variables have uncertainties", *American Journal of Physics* **52**, 22–26, 1984.

30. W.W. Morgan, "The MK system and the MK process", in *The MK Process and Stellar Classification*, ed. R.F. Garrison, David Dunlap Observatory, 18, 1984.

31. F. Murtagh, "Structure of hierarchic clusterings: implications for information retrieval and for multivariate data analysis", *Information Processing and Management*, **20**, 611–617, 1984.

32. F. Murtagh, *Multidimensional Clustering Algorithms*, COMPSTAT Lectures Volume 4, Physica-Verlag, Wien, 1985.

33. J.V. Narlikar, "Statistical techniques in astronomy", *Sankhya: The Indian Journal of Statistics*, Series B, Part 2, **44**, 125–134, 1982.

34. S. Okamura, "Global structure of Virgo cluster galaxies", in eds. O.G. Richter and B. Binggeli, *ESO Workshop on The Virgo Cluster of Galaxies*, ESO Conference and Workshop Proceedings No. 20, 201–215, 1985.

35. W.D. Pence and E. Davoust, "Supplement to the detailed bibliography on the surface photometry of galaxies", *Astronomy and Astrophysics Supplement Series*, **60**, 517–526, 1985.

36. J. Pfleiderer, "Data set description and bias", in *Statistical Methods in Astronomy*, European Space Agency Special Publication 201, 3–11, 1983.

37. A. Rapoport and S. Fillenbaum, "An experimental study of semantic structures", in eds. A.K. Romney, R.N. Shepard and S.B. Nerlove, *Multidimensional Scaling; Theory and Applications in the Behavioral Sciences. Vol. 2, Applications*, Seminar Press, New York, 93–131, 1972.

38. F.J. Rohlf, "Generalization of the gap test for the detection of multivariate outliers", *Biometrics*, **31**, 93–101, 1975.

39. G. Salton and M.J. McGill, *Introduction to Modern Information Retrieval*, McGraw-Hill, New York, 1983.

40. D. Sankoff and J.B. Kruskal (Eds.), *Time Warps, String Edits, and Macromolecules: The Theory and Practice of Sequence Comparison*, Addison-Wesley, New York, 1983.

41. M.J. Seaton, "Interstellar extinction in the UV", *Monthly Notices of the Royal Astronomical Society* **187**, 73P–76P, 1979.

42. B.T. Smith et al., *Matrix Eigensystem Routines — EISPACK Guide*, Lecture Notes in Computer Science 6, Springer Verlag, Berlin and New York, 1976.

43. P.H.A. Sneath and R.R. Sokal, *Numerical Taxonomy*, Freeman, San Francisco, 1973.

44. H. Späth, *Cluster Dissection and Analysis: Theory, Fortran Programs, Examples*, Ellis Horwood, Chichester, 1985.

45. B. Takase, K. Kodaira and S. Okamura, *An Atlas of Selected Galaxies*, University of Tokyo Press, Tokyo, 1984.

46. M. Thonnat, "Automatic morphological description of galaxies and classification by an expert system", INRIA Rapport de Recherche, Centre Sophia Antipolis, No. 387, 1985.

47. A. Tucker, *Applied Combinatorics*, Wiley, New York, 1980.

48. M. Volle, *Analyse des Données*, Economica, Paris, 1981.

49. J.V. Wall, "Practical statistics for astronomers. I. Definitions, the normal distribution, detection of signal", *Quarterly Journal of the Royal Astronomical Society*, **20**, 130–152, 1979.

50. M. Watanabe, K. Kodaira and S. Okamura, "Digital surface photometry of galaxies toward a quantitative classification. I. 20 galaxies in the Virgo cluster", *Astrophysics Journal Supplement Series.*, **50**, 1–22, 1982.

51. G.B. Wetherill, *Elementary Statistical Methods*, Chapman and Hall, London, 1972.

52. D. Wishart, "Mode analysis: a generalization of nearest neighbour which reduces chaining effects", in ed. A.J. Cole, *Numerical Taxonomy*, Academic Press, New York, 282–311, 1969.

53. D. York, "Least squares fitting of a straight line", *Canadian Journal of Physics* **44**, 1079–1086, 1966.

54. C.T. Zahn, "Graph–theoretical methods for detecting and describing Gestalt clusters", *IEEE Transactions on Computers*, **C–20**, 68–86, 1971.

Lelièvre, G. 11, 201
Lemaire, J. 125
Lentes, F.T. 30, 31
Leo (cluster) 81
Light curves 81
Lillie, C.F. 31
Linear combinations 26
Linear discriminant analysis 115
Linear regression 167
Lohnes, P.R. 172
Lunar surface 77, 78
Lutz, T.E. 170, 201
Lybanon, M. 168, 171, 201
Maccarone, M.C. 79, 80
Maceroni, C. 30
MacGillivray, H.T. 124
Machine vision 56
Maggi, P. 123
Mahalanobis distance 115
Malagnini, M.L. 124
Malinvaud, E. 172
Manzotti, G. 31
Maravalle, M. 30
Marriott, F.H.C. 33
Martin, R. 124
Mass (of a point) 157
Massa, D. 31
Massaro, E. 78
Materne, J. 81
Mathez 11, 201
Maximum likelihood discrimination 119
McCheyne, R.S. 79
McGill, M.J. 73, 85, 202
Meadows, A.J. 79
Median method 63
Melsa, J.L. 126
Mennessier, M.O. 81, 82, 170
Mersch, G. 80, 170
Metric scaling 163
Metric space 3
MIDAS (software) 81, 82

Minimal spanning tree 61
Minimum variance method 57
Minkowski distance 4
Mira type variables 82
Misclassification rate 121
Missing values 6
MK classification 123, 170, 174
Modalities (response categories) 160
Moles, M. 82
Montgomery, D.C. 171
Morgan, W.W. 175, 201
Morineau, A. 33, 84, 172, 201
Multidimensional scaling 163
Multiple correspondence analysis 160
Multiple discriminant analysis 113
Multiple regression 168
Murtagh, F. 65, 69, 82, 84, 201
Narlikar, J.V. 11, 201
Nearest neighbour chain 65
Nicolet, B. 32
Nicoll, J.F. 170, 171
Nobelis, Ph. 31, 80
Non-hierarchical clustering 56
Non-metric scaling 166
Normal galaxies 30
Normalization 5
Norm 17
Object classification 8, 78, 81, 123, 124
Objects 1
Okamura, S. 10, 11, 12, 32, 202, 203
Olmo, A. del 82
Operations research 7
Ordinal data 2
Orthogonality 18
Outliers 73
Overlapping clustering 75
Paolicchi, P. 30
Parallax solutions 31, 32
Parameters 1
Partitioning methods 57

INDEX

Galaxy luminosity profiles 82
Galaxy morphology 9
Galaxy photometry 9, 32
Galeotti, P. 31
Gaussian density function 118
Gavrishin, A.I. 77, 78
Geller, M.J. 81
Generalized distance 115
Geneva photometric system 81
Giovannelli, F. 83
Globular clusters 30
Gnanadesikan, R. 33
Gordon, A.D. 84, 166, 172, 200
Gott III, J.R. 169
Gower similarity 7
Gower, J.C. 7, 11, 163, 200
Graham, R.L. 84
Graph theory 65, 71
Greedy algorithm 71
Green, S.F. 79
Greenacre, M. 172
GSSS (software) 124
Guibert, J. 30
Hall, P.A.V. 8, 11, 200
Hamming distance 4
Hand, D.J. 119, 121, 125, 200
Hart, P. 125
Hartigan, J.A. 84
Heck, A. 31, 80, 123, 169, 170, 173, 174, 175, 182, 200
Hell, P. 84
Henry, R. 78
Hercules supercluster 82
Hierarchical cluster analysis 56
Hoffman, R.L. 79
Horseshoe pattern 162
Huchra, J.P. 81
Huyghen's theorem 70, 113
Image cataloging 123
IMSL (software) 125
Inertia 157

Information retrieval 68
Interferograms 83
Inversion (cluster analysis) 66
ISODATA (software) 77
Isotonic regression 170
Iterative optimization 75
IUE spectra 31, 80, 174
IUE 173
Jaccard similarity 7
Jambu, M. 84
James, M. 125
Janes, K.A. 30
Jarvis, J.F. 81, 123, 124
Jaschek, C. 11, 175, 200
Jaschek, M. 175, 201
Jasniewicz, G. 81
Jefferys, W.H. 170
K-NNs discrimination 121
Karhunen-Loève expansion 22
Kendall, M.G. 33, 126
Kerridge, S.J. 31, 32
Kodaira, K. 11, 12, 32, 202, 203
Kondo, Y. 174, 201
Koorneef, J. 31
Kruskal, J.B. 8, 11, 166, 172, 201, 202
Kruszewski, A. 81
Kuhn, J.R. 170
Kurtz, M.J. 8, 11, 31, 81, 123, 201
Lachenbruch, P.A. 126
Lagrangian multiplier 20
Lance-Williams formula 63
Large Magellanic Cloud 80
Lasota, J.P. 83
Lauberts, A. 9, 82
Least squares methods 169, 170, 171
Leaving one out method 122
Lebart, L. 33, 84, 172, 201
Lebeaux, M.O. 84
Lee, R.C.T. 84
Lefèvre, O. 9, 10, 11, 201
Legendre, P. 7, 11, 200

Collins, A.J. 33, 125
Colomba, G. 78
Complete disjunctive coding 160
Complete linkage method 61
Conditional probability 117
Confusion matrix 122
Contingency tables 156
Cooley, W.W. 172
Coradini, A. 77, 78, 83
Correlation coefficient 18
Correlations (PCA of) 24
Correspondence analysis 156
Cosmic rays 80, 83
COSMOS (software) 124
Covariance matrix, total 113
Covariances (PCA of) 24
Covariances, between classes 113
Covariances, within classes 113
Cowley, C.R. 78, 79, 169
Crézé, M. 169
Cucchiaro, A. 174, 200
Davies, J.K. 79
Davoust, E. 9, 10, 11, 200, 202
De Biase, G.A. 79
Decision theory 126
Deeming, T.J. 30, 169
Defays, D. 80
Dekker, H. 83
Dendrogram 59
Design set 111
Devijver, P.A. 79
Deville, J.C. 172
Di Gesù, V. 78, 79, 80
Diday, E. 125
Dimensionality reduction 26
Discriminant analysis 111
Discriminant factor analysis 113
Dissimilarity 3
Distance scales 170
Distance 3, 18
Doradus (cluster) 31

Doubling 162
Dowling, G.R. 8, 11, 200
Draper, N.R. 171
Dual spaces 21
Dubes, R.C. 79
Duda, R. 125
Dunn, G. 171
Dynamic programming 7
Eaton, N. 79
Efstathiou, G. 30
Egret, D. 31, 80
Eichhorn, H. 169
Eigenvalues 20, 22, 115, 159, 166
Eigenvectors 20, 159
EISPACK (software) 33
Elliptical galaxies 30, 32
Error rate 122
Estimation theory 126
Euclidean distance 3
Euclidean space 18
Everitt, B. 7, 10, 83, 171, 200
Exchange method 76
Extended objects 79, 80
Faber, S.M. 30
Factors 21, 160
Fall, S.M. 30
Feature selection 26, 122
Feitzinger, J.V. 80
Fillenbaum, S. 61, 84, 202
Fisher's linear discriminant 116
Fisher, R.A. 125
FOCAS (software) 81, 123
Fofi, M. 30
Fracassini, M. 31, 123, 169
Frank, I.E. 80
Fresneau, A. 80
Fukunaga, K. 125
Fulchignoni, M. 77, 78
Galactic rotation 169, 170
Galaxy clustering 77, 81
Galaxy evolution 124

Index

χ^2 distance 156
γ-ray astronomy 32, 79
A posteriori probability 117
A priori probability 117
Adorf, H.-M. 123, 199
Agglomerative algorithm 59
Aikman, G.C.L. 169
Albert, A. 80
Aldenderfer, M.S. 59, 83, 199
Alexander, L.W.G. 124
Algorithme des célibataires 66
Am, Ap stars 78
Anderberg, M.R. 7, 10, 77, 83, 199
Anderson, T.W. 33
Antonello, E. 31, 123, 169
Assignment 115
Asteroids 32, 78, 79, 82
Automatic classification 55
Average link method 61
Balkowski, C. 30
Barrow, J.D. 77
Barycentre 115
Bates, B.A. 80
Bayes' theorem 117
Bayesian discrimination 116
Benzécri, J.P. 83, 171
Bevington, P.R. 171, 199
Bhavsar, S.P. 77
Bianchi, R. 77
Bijaoui, A. 11, 29, 78, 201
Binary data 4
Binary stars 30
Blashfield, R.K. 59, 83, 199
Bock, H.H. 83

Boggess, A. 174, 199
Bow, S.-T. 125
Branham Jr., R.L. 169
Braunsfurth, E. 80
Brosche, P. 30
Brownlee, D.E. 80
Bubble chamber tracks 72
Buccheri, R. 78
Bujarrabal, V. 30
Burt table 162
Buser, R. 30, 169
Butchins, S.A. 78
Butler, J.C. 77
Cancer (cluster) 82
Canonical correlation analysis 167
Canonical correlations 167
Canonical discrimination 113
Canonical variate analysis 167
Carusi, A. 78
Categorical data 2, 156
Centring 5, 15
Centroid method 63
Cepheids 31, 169
Chaining (cluster analysis) 59
Chambers, J.M. 25, 33, 200
Chatfield, C. 33, 125
Chebyshev distance 5
Christian, C.A. 30
Chronometric cosmology 171
Classification 112
CLUSTAN (software) 83
Cluster analysis 55
Coffaro, P. 78
Cohn, D.L. 126

INDEX

Pasian, F. 124
Pasinetti, L.E. 31, 123, 169
Paturel, G. 82
Peek, E.A. 171
Pelat, D. 32
Pence, W.D. 9, 10, 11, 200, 202
Percentage inertia 155
Percentage variance 22
Perea, J. 82
Perseus supercluster 82
Peterson, D.M. 170
Pfleiderer, J. 2, 11, 202
Photometric systems 30, 31, 32, 81, 123, 169
Picat, J.P. 11, 201
Pirenne, B. 83
Planetary surfaces 77
Plug in estimators 119
Polimene, M.L. 83
Ponz, D. 83
Pouget, J. 125
Pratt, N.M. 124
Principal components analysis 13
Principal coordinates analysis 163
Profiles 156
Projection 18
Proper motions 170
Pucillo, M. 124
Quadratic discrimination 116
Quadratic form 20
Qualitative data 2
Quantitative data 2
Quasars 171
Radial velocities 169
Raffaelli, G. 31, 123
Rapoport, A. 61, 84, 202
Reciprocal nearest neighbours 66
Reddening (interstellar) 31
Reddish, V.C. 124
Redshift 170
Reversal (cluster analysis) 66

Reynolds, M.L. 172
RGU photometry 30, 169
Rohlf, F.J. 73, 84, 202
Romeder, J.M. 126
Sacco, B. 79
SAI (software) 30
Salemi, S. 78
Salton, G. 73, 85, 202
Sankoff, D. 8, 11, 202
Santin, P. 124
SAS (software) 126
Scalar product 17
Scaling of data 3
Schifman, S.S. 172
Seaton, M.J. 178, 202
Seber, G.A.F. 171
Sebok, W.J. 124
Seddon, H. 124
Segal, I.E. 170, 171
Seriation 163
Signal detection 11
Similarity 3
Single linkage method 57
Smith, B.T. 25, 33, 202
Smith, H. 171
Sneath, P.H.A. 59, 85, 202
Sokal, R.R. 59, 85, 202
Solar motion 169
Sonoda, D.H. 77
Spark chambers 78
Späth, H. 77, 85, 202
Spectral classification 30, 31, 32, 80, 123, 170, 175
Spectroscopic binaries 30
Spheroidal galaxies 31
Spiral galaxies 31, 32
Standardization 5, 23
Standish Jr., M. 169
Stellar abundances 169
Stellar classification 78, 80, 175
Stellar distances 169

Stellar spectra 79, 175
Stepwise regression 170
Strong, A.W. 32
Sums of squares/cross products 20
Supervised classification 112
Supplementary rows, columns 160
Takase, B. 10, 11, 32, 202
TD1 (satellite) 174
Testu, F. 125
Tholen, D.J. 32, 82
Thonnat, M. 9, 11, 202
Tobia, G. 79
Torgerson, W.S. 163, 172
Training set 111
Transition formulas 160
Triangular inequality 4
Tucker, A. 72, 85, 202
Turlot, J.C. 31, 80
Turner, E.L. 169
Types of data: categorical data 156
Types of data: contingency tables 156
Types of data: frequencies 156
Types of data: mixed data 156
Types of data: probabilities 156
Tyson, J.A. 81, 123, 124
UBV photometry 30, 169
Unsupervised classification 112
Upgren, A.R. 31, 32
UV spectra 80
uvby photometry 30
uvbyβ photometry 170
Vader, J.P. 32
Valdes, F. 124
Variable stars 81, 123
Variables 1
Variance–covariance matrix, total 113
Virgo (cluster) 12, 32, 82
Walker, G.S. 124
Wall, J.V. 2, 11, 203
Ward's method 57
Warwick, K.M. 33, 84, 172, 201

Watanabe, M. 9, 12, 32, 203
Weight (of a point) 157
Wetherill, G.B. 2, 12, 171, 203
Whitmore, B.C. 32
Whitney, C.A. 32
Williams, P.R. 124
Wish, M. 166, 172, 201
Wishart, D. 68, 85, 203
York, D. 168, 171, 203
Young, F.L. 172
Zahn, C.T. 72, 85, 203
Zandonella, A. 83

ASTROPHYSICS AND SPACE SCIENCE LIBRARY

Edited by

J. E. Blamont, R. L. F. Boyd, L. Goldberg, C. de Jager, Z. Kopal, G. H. Ludwig, R. Lüst,
B. M. McCormac, H. E. Newell, L. I. Sedov, Z. Švestka

1. C. de Jager (ed.), *The Solar Spectrum, Proceedings of the Symposium held at the University of Utrecht, 26–31 August, 1963.* 1965, XIV + 417 pp.
2. J. Orthner and H. Maseland (eds.), *Introduction to Solar Terrestrial Relations, Proceedings of the Summer School in Space Physics held in Alpbach, Austria, July 15–August 10, 1963 and Organized by the European Preparatory Commission for Space Research.* 1965, IX + 506 pp.
3. C. C. Chang and S. S. Huang (eds.), *Proceedings of the Plasma Space Science Symposium, held at the Catholic University of America, Washington, D.C., June 11–14, 1963.* 1965, IX + 377 pp.
4. Zdeněk Kopal, *An Introduction to the Study of the Moon.* 1966, XII + 464 pp.
5. B. M. McCormac (ed.), *Radiation Trapped in the Earth's Magnetic Field. Proceedings of the Advanced Study Institute, held at the Chr. Michelsen Institute, Bergen, Norway, August 16–September 3, 1965.* 1966, XII + 901 pp.
6. A. B. Underhill, *The Early Type Stars.* 1966, XII + 282 pp.
7. Jean Kovalevsky, *Introduction to Celestial Mechanics.* 1967, VIII + 427 pp.
8. Zdeněk Kopal and Constantine L. Goudas (eds.), *Measure of the Moon. Proceedings of the 2nd International Conference on Selenodesy and Lunar Topography, held in the University of Manchester, England, May 30–June 4, 1966.* 1967, XVIII + 479 pp.
9. J. G. Emming (ed.), *Electromagnetic Radiation in Space. Proceedings of the 3rd ESRO Summer School in Space Physics, held in Alpbach, Austria, from 19 July to 13 August, 1965.* 1968, VIII + 307 pp.
10. R. L. Carovillano, John F. McClay, and Henry R. Radoski (eds.), *Physics of the Magnetosphere, Based upon the Proceedings of the Conference held at Boston College, June 19–28, 1967.* 1968, X + 686 pp.
11. Syun-Ichi Akasofu, *Polar and Magnetospheric Substorms.* 1968, XVIII + 280 pp.
12. Peter M. Millman (ed.), *Meteorite Research. Proceedings of a Symposium on Meteorite Research, held in Vienna, Austria, 7–13 August, 1968.* 1969, XV + 941 pp.
13. Margherita Hack (ed.), *Mass Loss from Stars. Proceedings of the 2nd Trieste Colloquium on Astrophysics, 12–17 September, 1968.* 1969, XII + 345 pp.
14. N. D'Angelo (ed.), *Low-Frequency Waves and Irregularities in the Ionosphere. Proceedings of the 2nd ESRIN-ESLAB Symposium, held in Frascati, Italy, 23–27 September, 1968.* 1969, VII + 218 pp.
15. G. A. Partel (ed.), *Space Engineering. Proceedings of the 2nd International Conference on Space Engineering, held at the Fondazione Giorgio Cini, Isola di San Giorgio, Venice, Italy, May 7–10, 1969.* 1970, XI + 728 pp.
16. S. Fred Singer (ed.), *Manned Laboratories in Space. Second International Orbital Laboratory Symposium.* 1969, XIII + 133 pp.
17. B. M. McCormac (ed.), *Particles and Fields in the Magnetosphere. Symposium Organized by the Summer Advanced Study Institute, held at the University of California, Santa Barbara, Calif., August 4–15, 1969.* 1970, XI + 450 pp.
18. Jean-Claude Pecker, *Experimental Astronomy.* 1970, X + 105 pp.
19. V. Manno and D. E. Page (eds.), *Intercorrelated Satellite Observations related to Solar Events. Proceedings of the 3rd ESLAB/ESRIN Symposium held in Noordwijk, The Netherlands, September 16–19, 1969.* 1970, XVI + 627 pp.
20. L. Mansinha, D. E. Smylie, and A. E. Beck, *Earthquake Displacement Fields and the Rotation of the Earth, A NATO Advanced Study Institute Conference Organized by the Department of Geophysics, University of Western Ontario, London, Canada, June 22–28, 1969.* 1970, XI + 308 pp.
21. Jean-Claude Pecker, *Space Observatories.* 1970, XI + 120 pp.
22. L. N. Mavridis (ed.), *Structure and Evolution of the Galaxy. Proceedings of the NATO Advanced Study Institute, held in Athens, September 8–19, 1969.* 1971, VII + 312 pp.

23. A. Muller (ed.), *The Magellanic Clouds. A European Southern Observatory Presentation: Principal Prospects, Current Observational and Theoretical Approaches, and Prospects for Future Research*, Based on the Symposium on the Magellanic Clouds, held in Santiago de Chile, March 1969, on the Occasion of the Dedication of the European Southern Observatory. 1971, XII + 189 pp.
24. B. M. McCormac (ed.), *The Radiating Atmosphere*. Proceedings of a Symposium Organized by the Summer Advanced Study Institute, held at Queen's University, Kingston, Ontario, August 3–14, 1970. 1971, XI + 455 pp.
25. G. Fiocco (ed.), *Mesospheric Models and Related Experiments*. Proceedings of the 4th ESRIN-ESLAB Symposium, held at Frascati, Italy, July 6–10, 1970. 1971, VIII + 298 pp.
26. I. Atanasijević, *Selected Exercises in Galactic Astronomy*. 1971, XII + 144 pp.
27. C. J. Macris (ed.), *Physics of the Solar Corona*. Proceedings of the NATO Advanced Study Institute on Physics of the Solar Corona, held at Cavouri-Vouliagmeni, Athens, Greece, 6–17 September 1970. 1971, XII + 345 pp.
28. F. Delobeau, *The Environment of the Earth*. 1971, IX + 113 pp.
29. E. R. Dyer (general ed.), *Solar-Terrestrial Physics/1970*. Proceedings of the International Symposium on Solar-Terrestrial Physics, held in Leningrad, U.S.S.R., 12–19 May 1970. 1972, VIII + 938 pp.
30. V. Manno and J. Ring (eds.), *Infrared Detection Techniques for Space Research*. Proceedings of the 5th ESLAB-ESRIN Symposium, held in Noordwijk, The Netherlands, June 8–11, 1971. 1972, XII + 344 pp.
31. M. Lecar (ed.), *Gravitational N-Body Problem*. Proceedings of IAU Colloquium No. 10, held in Cambridge, England, August 12–15, 1970. 1972, XI + 441 pp.
32. B. M. McCormac (ed.), *Earth's Magnetospheric Processes*. Proceedings of a Symposium Organized by the Summer Advanced Study Institute and Ninth ESRO Summer School, held in Cortina, Italy, August 30–September 10, 1971. 1972, VIII + 417 pp.
33. Antonin Rükl, *Maps of Lunar Hemispheres*. 1972, V + 24 pp.
34. V. Kourganoff, *Introduction to the Physics of Stellar Interiors*. 1973, XI + 115 pp.
35. B. M. McCormac (ed.), *Physics and Chemistry of Upper Atmospheres*. Proceedings of a Symposium Organized by the Summer Advanced Study Institute, held at the University of Orléans, France, July 31–August 11, 1972. 1973, VIII + 389 pp.
36. J. D. Fernie (ed.), *Variable Stars in Globular Clusters and in Related Systems*. Proceedings of the IAU Colloquium No. 21, held at the University of Toronto, Toronto, Canada, August 29–31, 1972. 1973, IX + 234 pp.
37. R. J. L. Grard (ed.), *Photon and Particle Interaction with Surfaces in Space*. Proceedings of the 6th ESLAB Symposium, held at Noordwijk, The Netherlands, 26–29 September, 1972. 1973, XV + 577 pp.
38. Werner Israel (ed.), *Relativity, Astrophysics and Cosmology*. Proceedings of the Summer School, held 14–26 August 1972, at the BANFF Centre, BANFF, Alberta, Canada. 1973, IX + 323 pp.
39. B. D. Tapley and V. Szebehely (eds.), *Recent Advances in Dynamical Astronomy*. Proceedings of the NATO Advanced Study Institute in Dynamical Astronomy, held in Cortina d'Ampezzo, Italy, August 9–12, 1972. 1973, XIII + 468 pp.
40. A. G. W. Cameron (ed.), *Cosmochemistry*. Proceedings of the Symposium on Cosmochemistry, held at the Smithsonian Astrophysical Observatory, Cambridge, Mass., August 14–16, 1972. 1973, X + 173 pp.
41. M. Golay, *Introduction to Astronomical Photometry*. 1974, IX + 364 pp.
42. D. E. Page (ed.), *Correlated Interplanetary and Magnetospheric Observations*. Proceedings of the 7th ESLAB Symposium, held at Saulgau, W. Germany, 22–25 May, 1973. 1974, XIV + 662 pp.
43. Riccardo Giacconi and Herbert Gursky (eds.), *X-Ray Astronomy*. 1974, X + 450 pp.
44. B. M. McCormac (ed.), *Magnetospheric Physics*. Proceedings of the Advanced Summer Institute, held in Sheffield, U.K., August 1973. 1974, VII + 399 pp.
45. C. B. Cosmovici (ed.), *Supernovae and Supernova Remnants*. Proceedings of the International Conference on Supernovae, held in Lecce, Italy, May 7–11, 1973. 1974, XVII + 387 pp.
46. A. P. Mitra, *Ionospheric Effects of Solar Flares*. 1974, XI + 294 pp.
47. S.-I. Akasofu, *Physics of Magnetospheric Substorms*. 1977, XVIII + 599 pp.

48. H. Gursky and R. Ruffini (eds.), *Neutron Stars, Black Holes and Binary X-Ray Sources.* 1975, XII + 441 pp.
49. Z. Švestka and P. Simon (eds.), *Catalog of Solar Particle Events 1955–1969. Prepared under the Auspices of Working Group 2 of the Inter-Union Commission on Solar-Terrestrial Physics.* 1975, IX + 428 pp.
50. Zdeněk Kopal and Robert W. Carder, *Mapping of the Moon.* 1974, VIII + 237 pp.
51. B. M. McCormac (ed.), *Atmospheres of Earth and the Planets. Proceedings of the Summer Advanced Study Institute, held at the University of Liège, Belgium, July 29–August 8, 1974.* 1975, VII + 454 pp.
52. V. Formisano (ed.), *The Magnetospheres of the Earth and Jupiter. Proceedings of the Neil Brice Memorial Symposium, held in Frascati, May 28–June 1, 1974.* 1975, XI + 485 pp.
53. R. Grant Athay, *The Solar Chromosphere and Corona: Quiet Sun.* 1976, XI + 504 pp.
54. C. de Jager and H. Nieuwenhuijzen (eds.), *Image Processing Techniques in Astronomy. Proceedings of a Conference, held in Utrecht on March 25–27, 1975.* 1976, XI + 418 pp.
55. N. C. Wickramasinghe and D. J. Morgan (eds.), *Solid State Astrophysics. Proceedings of a Symposium, held at the University College, Cardiff, Wales, 9–12 July, 1974.* 1976, XII + 314 pp.
56. John Meaburn, *Detection and Spectrometry of Faint Light.* 1976, IX + 270 pp.
57. K. Knott and B. Battrick (eds.), *The Scientific Satellite Programme during the International Magnetospheric Study. Proceedings of the 10th ESLAB Symposium, held at Vienna, Austria, 10–13 June 1975.* 1976, XV + 464 pp.
58. B. M. McCormac (ed.), *Magnetospheric Particles and Fields. Proceedings of the Summer Advanced Study School, held in Graz, Austria, August 4–15, 1975.* 1976, VII + 331 pp.
59. B. S. P. Shen and M. Merker (eds.), *Spallation Nuclear Reactions and Their Applications.* 1976, VIII + 235 pp.
60. Walter S. Fitch (ed.), *Multiple Periodic Variable Stars. Proceedings of the International Astronomical Union Colloquium No. 29, held at Budapest, Hungary, 1–5 September 1976.* 1976, XIV + 348 pp.
61. J. J. Burger, A. Pedersen, and B. Battrick (eds.), *Atmospheric Physics from Spacelab. Proceedings of the 11th ESLAB Symposium, Organized by the Space Science Department of the European Space Agency, held at Frascati, Italy, 11–14 May 1976.* 1976, XX + 409 pp.
62. J. Derral Mulholland (ed.), *Scientific Applications of Lunar Laser Ranging. Proceedings of a Symposium held in Austin, Tex., U.S.A., 8–10 June, 1976.* 1977, XVII + 302 pp.
63. Giovanni G. Fazio (ed.), *Infrared and Submillimeter Astronomy. Proceedings of a Symposium held in Philadelphia, Penn., U.S.A., 8–10 June, 1976.* 1977, X + 226 pp.
64. C. Jaschek and G. A. Wilkins (eds.), *Compilation, Critical Evaluation and Distribution of Stellar Data. Proceedings of the International Astronomical Union Colloquium No. 35, held at Strasbourg, France, 19–21 August, 1976.* 1977, XIV + 316 pp.
65. M. Friedjung (ed.), *Novae and Related Stars. Proceedings of an International Conference held by the Institut d'Astrophysique, Paris, France, 7–9 September, 1976.* 1977, XIV + 228 pp.
66. David N. Schramm (ed.), *Supernovae. Proceedings of a Special IAU-Session on Supernovae held in Grenoble, France, 1 September, 1976.* 1977, X + 192 pp.
67. Jean Audouze (ed.), *CNO Isotopes in Astrophysics. Proceedings of a Special IAU Session held in Grenoble, France, 30 August, 1976.* 1977, XIII + 195 pp.
68. Z. Kopal, *Dynamics of Close Binary Systems.* XIII + 510 pp.
69. A. Bruzek and C. J. Durrant (eds.), *Illustrated Glossary for Solar and Solar-Terrestrial Physics.* 1977, XVIII + 204 pp.
70. H. van Woerden (ed.), *Topics in Interstellar Matter.* 1977, VIII + 295 pp.
71. M. A. Shea, D. F. Smart, and T. S. Wu (eds.), *Study of Travelling Interplanetary Phenomena.* 1977, XII + 439 pp.
72. V. Szebehely (ed.), *Dynamics of Planets and Satellites and Theories of Their Motion. Proceedings of IAU Colloquium No. 41, held in Cambridge, England, 17–19 August 1976.* 1978, XII + 375 pp.
73. James R. Wertz (ed.), *Spacecraft Attitude Determination and Control.* 1978, XVI + 858 pp.

74. Peter J. Palmadesso and K. Papadopoulos (eds.), *Wave Instabilities in Space Plasmas. Proceedings of a Symposium Organized Within the XIX URSI General Assembly held in Helsinki, Finland, July 31–August 8, 1978.* 1979, VII + 309 pp.
75. Bengt E. Westerlund (ed.), *Stars and Star Systems. Proceedings of the Fourth European Regional Meeting in Astronomy held in Uppsala, Sweden, 7–12 August, 1978.* 1979, XVIII + 264 pp.
76. Cornelis van Schooneveld (ed.), *Image Formation from Coherence Functions in Astronomy. Proceedings of IAU Colloquium No. 49 on the Formation of Images from Spatial Coherence Functions in Astronomy, held at Groningen, The Netherlands, 10–12 August 1978.* 1979, XII + 338 pp.
77. Zdeněk Kopal, *Language of the Stars. A Discourse on the Theory of the Light Changes of Eclipsing Variables.* 1979, VIII + 280 pp.
78. S.-I. Akasofu (ed.), *Dynamics of the Magnetosphere. Proceedings of the A.G.U. Chapman Conference 'Magnetospheric Substorms and Related Plasma Processes' held at Los Alamos Scientific Laboratory, N.M., U.S.A., October 9–13, 1978.* 1980, XII + 658 pp.
79. Paul S. Wesson, *Gravity, Particles, and Astrophysics. A Review of Modern Theories of Gravity and G-variability, and their Relation to Elementary Particle Physics and Astrophysics.* 1980, VIII + 188 pp.
80. Peter A. Shaver (ed.), *Radio Recombination Lines. Proceedings of a Workshop held in Ottawa, Ontario, Canada, August 24–25, 1979.* 1980, X + 284 pp.
81. Pier Luigi Bernacca and Remo Ruffini (eds.), *Astrophysics from Spacelab.* 1980, XI + 664 pp.
82. Hannes Alfvén, *Cosmic Plasma*, 1981, X + 160 pp.
83. Michael D. Papagiannis (ed.), *Strategies for the Search for Life in the Universe*, 1980, XVI + 254 pp.
84. H. Kikuchi (ed.), *Relation between Laboratory and Space Plasmas*, 1981, XII + 386 pp.
85. Peter van der Kamp, *Stellar Paths*, 1981, XXII + 155 pp.
86. E. M. Gaposchkin and B. Kołaczek (eds.), *Reference Coordinate Systems for Earth Dynamics*, 1981, XIV + 396 pp.
87. R. Giacconi (ed.), *X-Ray Astronomy with the Einstein Satellite. Proceedings of the High Energy Astrophysics Division of the American Astronomical Society Meeting on X-Ray Astronomy held at the Harvard-Smithsonian Center for Astrophysics, Cambridge, Mass., U.S.A., January 28–30, 1980.* 1981, VII + 330 pp.
88. Icko Iben Jr. and Alvio Renzini (eds.), *Physical Processes in Red Giants. Proceedings of the Second Workshop, helt at the Ettore Majorana Centre for Scientific Culture, Advanced School of Agronomy, in Erice, Sicily, Italy, September 3–13, 1980.* 1981, XV + 488 pp.
89. C. Chiosi and R. Stalio (eds.), *Effect of Mass Loss on Stellar Evolution. IAU Colloquium No. 59 held in Miramare, Trieste, Italy, September 15–19, 1980.* 1981, XXII + 532 pp.
90. C. Goudis, *The Orion Complex: A Case Study of Interstellar Matter.* 1982, XIV + 306 pp.
91. F. D. Kahn (ed.), *Investigating the Universe. Papers Presented to Zdenek Kopal on the Occasion of his retirement, September 1981.* 1981, X + 458 pp.
92. C. M. Humphries (ed.), *Instrumentation for Astronomy with Large Optical Telescopes, Proceedings of IAU Colloquium No. 67.* 1981, XVII + 321 pp.
93. R. S. Roger and P. E. Dewdney (eds.), *Regions of Recent Star Formation, Proceedings of the Symposium on "Neutral Clouds Near HII Regions - Dynamics and Photochemistry", held in Penticton, B.C., June 24–26, 1981.* 1982, XVI + 496 pp.
94. O. Calame (ed.), *High-Precision Earth Rotation and Earth-Moon Dynamics. Lunar Distances and Related Observations.* 1982, XX + 354 pp.
95. M. Friedjung and R. Viotti (eds.), *The Nature of Symbiotic Stars*, 1982, XX + 310 pp.
96. W. Fricke and G. Teleki (eds.), *Sun and Planetary System*, 1982, XIV + 538 pp.
97. C. Jaschek and W. Heintz (eds.), *Automated Data Retrieval in Astronomy*, 1982, XX + 324 pp.
98. Z. Kopal and J. Rahe (eds.), *Binary and Multiple Stars as Tracers of Stellar Evolution*, 1982, XXX + 503 pp.
99. A. W. Wolfendale (ed.), *Progress in Cosmology*, 1982, VIII + 360 pp.
100. W. L. H. Shuter (ed.), *Kinematics, Dynamics and Structure of the Milky Way*, 1983, XII + 392 pp.

101. M. Livio and G. Shaviv (eds.), *Cataclysmic Variables and Related Objects*, 1983, XII + 351 pp.
102. P. B. Byrne and M. Rodonò (eds.), *Activity in Red-Dwarf Stars*, 1983, XXVI + 670 pp.
103. A. Ferrari and A. G. Pacholczyk (eds.), *Astrophysical Jets*, 1983, XVI + 328 pp.
104. R. L. Carovillano and J. M. Forbes (eds.), *Solar-Terrestrial Physics*, 1983, XVIII + 860 pp.
105. W. B. Burton and F. P. Israel (eds.), *Surveys of the Southern Galaxy*, 1983, XIV + 310 pp.
106. V. V. Markellos and Y. Kozai (eds.), *Dynamical Trapping and Evolution on the Solar System*, 1983, XVI + 424 pp.
107. S. R. Pottasch, *Planetary Nebulae*, 1984, X + 322 pp.
108. M. F. Kessler and J. P. Phillips (eds.), *Galactic and Extragalactic Infrared Spectroscopy*, 1984, XII + 472 pp.
109. C. Chiosi and A. Renzini (eds.), *Stellar Nucleosynthesis*, 1984, XIV + 398 pp.
110. M. Capaccioli (ed.), *Astronomy with Schmidt-type Telescopes*, 1984, XXII + 620 pp.
111. F. Mardirossian, G. Giuricin, and M. Mezzetti (eds.), *Clusters and Groups of Galaxies*, 1984, XXII + 659 pp.
112. L. H. Aller, *Physics of Thermal Gaseous Nebulae*, 1984, X + 350 pp.
113. D. Q. Lamb and J. Patterson (eds.), *Cataclysmic Variables and Low-Mass X-Ray Binaries*, 1985, XII + 452 pp.
114. M. Jaschek and P. C. Keenan (eds.), *Cool Stars with Excesses of Heavy Elements*, 1985, XVI + 398 pp.
115. A. Carusi and G. B. Valsecchi (eds.), *Dynamics of Comets: Their Origin and Evolution*, 1985, XII + 442 pp.
116. R. M. Hjellming and D. M. Gibson (eds.), *Radio Stars*, 1985, XI + 411 pp.
117. M. Morris and B. Zuckermann (eds.), *Mass Loss from Red Giants*, 1985, xvi +320 pp.
118. Y. Sekido and H. Elliot (eds.), *Early History of Cosmic Ray Studies*, 1985, xvi + 444 pp.
119. R. H. Giese and P. Lamy (eds.), *Properties and Interactions of Interplanetary Dust*, 1985, xxvi + 444 pp.
120. W. Boland and H. van Woerden (eds.), *Birth and Evolution of Massive Stars and Stellar Groups*, 1985, xiv + 377 pp.
121. G. Giuricin, F. Mardirossian, M. Mezzetti, and M. Ramella (eds.), *Structure and Evolution of Active Galactic Nuclei*, 1986, xxvi +772 pp.
122. C. Chiosi and A. Renzini (eds.), *Spectral Evolution of Galaxies*, 1986, xii + 490 pp.
123. R. G. Marsden (ed.), *The Sun and the Heliosphere in Three Dimensions*, 1986, xii +525 pp.
124. F. P. Israel (ed.), *Light on Dark Matter*, 1986, xxiv +541 pp.
125. C. R. Cowley, M. M. Dworetsky, and C. Mégessier (eds.), *Upper Main Sequence Stars with Anomalous Abundances*, 1986, xiv +489 pp.
126. Y. Kamide and J. A. Slavin (eds.), *Solar Wind-Magnetosphere Coupling*, 1986, xvi +807 pp.
127. K. B. Bhatnagar (ed.), *Space Synamics and Celestial Mechanics*, 1986, xxvi +458 pp.
128. K. Hunger, D. Schönberner, and N. Kameswara Rao (eds.), *Hydrogen Deficient Stars and Related Objects*, 1986, xvii + 506 pp.
129. Y. Kondo, W. Wamsteker et al. (eds.), *Exploring the Universe with the IUE Satellite*, 1987, forthcoming.
130. V. I. Ferronsky, S. A. Denisik, and S. V. Ferronsky, *Jacobi Dynamics*, 1987, xii +365 pp.
131. F. Murtagh and A. Heck, *Multivariate Data Analysis*, 1987, xvi +210 pp.